Intratheater Airlift Functional Solution Analysis (FSA)

David T. Orletsky, Daniel M. Norton,
Anthony D. Rosello, William Stanley,
Michael Kennedy, Michael Boito,
Brian G. Chow, Yool Kim

Prepared for the United States Air Force
Approved for public release; distribution unlimited

PROJECT AIR FORCE

The research described in this report was sponsored by the United States Air Force under Contract FA7014-06-C-0001. Further information may be obtained from the Strategic Planning Division, Directorate of Plans, Hq USAF.

Library of Congress Cataloging-in-Publication Data is available for this publication.

ISBN: 978-0-8330-5085-4

Published 2011 by the RAND Corporation
1776 Main Street, P.O. Box 2138, Santa Monica, CA 90407-2138
1200 South Hayes Street, Arlington, VA 22202-5050
4570 Fifth Avenue, Suite 600, Pittsburgh, PA 15213-2665
RAND URL: http://www.rand.org/
To order RAND documents or to obtain additional information, contact
Distribution Services: Telephone: (310) 451-7002;
Fax: (310) 451-6915; Email: order@rand.org

Preface

This functional solution analysis (FSA) for U.S. Air Force (USAF) intratheater airlift is the third in a series of three documents that together constitute a capabilities-based assessment (CBA) required as part of the Joint Capabilities Integration and Development System (JCIDS). The first document in the series, the functional area analysis (FAA),[1] identified the operational tasks, conditions, and standards needed to achieve military objectives—in this case, certain intratheater airlift missions. The second, the functional needs analysis (FNA),[2] assessed the ability of the current assets to deliver the capabilities identified in the FAA. The third document in the series, the FSA, is an operationally based assessment of current capabilities to determine whether a materiel solution is required to close any identified capability gap identified in the FNA. In this case, the FSA determined that no nonmateriel solution could address the shortfall. Therefore, an analysis of alternatives (AoA) was undertaken to evaluate the cost-effectiveness of various materiel solutions.

The intratheater airlift FNA found that, within the next decade or so, the available C-130 fleet would no longer be able to meet the minimum requirement identified in the Mobility Capabilities Study

---

[1] David T. Orletsky, Anthony D. Rosello, and John Stillion, *Intratheater Airlift Functional Area Analysis (FAA)*, Santa Monica, Calif.: RAND Corporation, MG-685-AF, 2011.

[2] John Stillion, David T. Orletsky, and Anthony D. Rosello, *Intratheater Airlift Functional Needs Analysis (FNA)*, Santa Monica, Calif.: RAND Corporation, MG-822-AF, 2011.

(MCS).[3] It was therefore necessary to undertake an FSA to determine whether a nonmateriel solution could close this capability gap.

This monograph presents the results of the FSA. The analysis considered a large number of potential nonmateriel solutions that included changes in doctrine, organization, training, leadership, personnel, and facilities. Because service-life extension programs (SLEPs) and new aircraft acquisitions are materiel solutions, this FSA touches on them only briefly, reserving detailed analysis for the AoA.

Maj Gen Thomas P. Kane, Director, Plans and Programs, Headquarters, Air Mobility Command, Scott Air Force Base, Illinois (Headquarters AMC/A5), sponsored this research. The work was performed within the Aerospace Force Development Program of RAND Project AIR FORCE as part of a fiscal year (FY) 2007 study, "Intratheater Cargo Delivery Functional Solution Analysis (FSA)."

RAND Project AIR FORCE

RAND Project AIR FORCE (PAF), a division of the RAND Corporation, is the U.S. Air Force's federally funded research and development center for studies and analyses. PAF provides the Air Force with independent analyses of policy alternatives affecting the development, employment, combat readiness, and support of current and future aerospace forces. Research is conducted in four programs: Force Modernization and Employment; Manpower, Personnel, and Training; Resource Management; and Strategy and Doctrine.

Additional information about PAF is available on our website: http://www.rand.org/paf/

---

[3] U.S. Department of Defense and the Joint Chiefs of Staff, *Mobility Capabilities Study*, Washington, D.C., December 2005, Not Available to the General Public.

Contents

Figures

Tables

Summary

This monograph reports the findings of the FSA that RAND Project AIR FORCE produced for USAF intratheater airlift. The FSA is the third in a series of analyses that together constitute a CBA required as part of the JCIDS. The first, the FAA, identified the operational tasks, conditions, and standards needed to achieve military objectives—in this case, certain intratheater airlift missions.[1] The second, the FNA, assessed the ability of the current assets to deliver the capabilities identified in the FAA. The third document in the series, this FSA, assesses changes to current operations to determine whether a nonmateriel solution could close the capability gap identified in the FNA. If the FSA is unable to identify a nonmateriel solution to address the shortfall, an AoA is then undertaken to evaluate the cost-effectiveness of various materiel solutions. In this case, the analysis conducted after the FSA was called the USAF Intratheater Airlift Fleet Mix Analysis (UIAFMA).[2]

This assessment focuses on the movement of intratheater cargo and personnel. This mission is primarily driven by the joint land force requirement to move personnel, equipment, and supplies throughout the battlespace.

---

[1]   Orletsky, Rosello, and Stillion, 2011.

[2]   Michael Kennedy, David T. Orletsky, Anthony D. Rosello, Sean Bednarz, Katherine Comanor, Paul Dreyer, Chris Fitzmartin, Ken Munson, William Stanley, and Fred Timson, *USAF Intratheater Airlift Fleet Mix Analysis*, Santa Monica, Calif.: RAND Corporation, 2010, Not Available to the General Public.

The FAA identified three broad operational mission areas for intratheater airlift: (1) routine sustainment, (2) time-sensitive, mission-critical (TS/MC) resupply, and (3) maneuver capabilities to U.S. and allied forces across all operating environments. The FAA drew the tasks, conditions, and standards required for intratheater airlift from a review of national strategy and official Department of Defense publications.

Two potential capability gaps were identified and analyzed in the FNA. The first is to maintain a sufficient number of C-130s to meet the requirement identified in the MCS.[3] The MCS set the minimum number of USAF mobility air forces (MAF) C-130s at 395 total aircraft inventory (TAI). A significant and growing portion of the C-130 fleet is either operating under flight restrictions or grounded because of fatigue-related cracking of key structural components of the center wing box (CWB). The second is to provide responsive intratheater resupply in support of the U.S. Army. The FNA looked at providing both routine sustainment and TS/MC resupply of a sizable multibrigade combat team (BCT) ground force. Although large multi-BCT forces operating without a ground line of communication is not the current Army concept for future operations, the trend is toward more-dispersed operations of ground forces. Future ground forces will rely on increased aerial distribution.[4]

The FNA found that if the policies of imposing flight restrictions and grounding aircraft remain in place and nothing else is done, then the number of unrestricted C-130s available to the USAF is projected to fall below the minimum threshold of 395 in the next several years. Further, the FNA found that routine sustainment of a ground combat force of moderate size by the existing intratheater airlift system is extremely challenging. In most of the cases analyzed, the number of C-130s required to supply six BCTs by air was at or beyond the number of C-130s likely to be available to support any one operation.[5] For

---

[3]   DoD and JCS, 2005.

[4]   See U.S. Army Aviation Center, Futures Development Division, Directorate of Combat Developments, *Army Fixed Wing Aviation Functional Needs Analysis Report*, Fort Rucker, Ala., June 23, 2003b, p. 16-17.

[5]   We assumed one aircraft delivery to each of 18 battalion locations every eight hours.

TS/MC resupply, the FNA found that the existing intratheater airlift assets can be combined to provide a robust, responsive system with a reasonably small commitment of resources. In addition, the analysis suggests that allocating additional resources to the TS/MC mission results in rapidly diminishing returns in terms of reduced transit time. Since routine resupply is not a requirement and since TS/MC resupply takes relatively few assets, the FNA determined that the FSA should focus on ensuring that the intratheater airlift fleet continues to meet the 395 C-130 requirement identified in the MCS.[6] This requirement needs to be met in light of the large number of aircraft that are expected to undergo flight restrictions and groundings during the next two decades. Using each aircraft's unique annual flying rate and equivalent baseline hours (EBH) accumulation rate, Figure S.1 projects the decline in the MAF inventory of C-130s over time as they reach the

Figure S.1
Number of C-130s in MAF Inventory and MCS Requirement

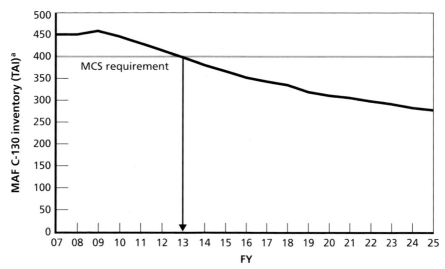

ᵃ Inventory numbers assume that all aircraft undergo TCTO 1908 inspection and are able to fly 45,000 EBH.
RAND MG818-S.1

---

6   DoD and JCS, 2005.

grounding limit.[7] The number of C-130s is projected to fall below the MCS requirement of 395 in 2013.[8]

Dealing with this emerging shortfall is complicated by the fact that aircraft are not distributed equally by age across active and reserve components. Figure S.2 shows that, in terms of years of service life remaining, the oldest aircraft are primarily in the active component, while the majority of the newer aircraft are in the reserve component. Thus, heavily tasked active forces face the most immediate prospect of not having enough aircraft to perform their missions.

FSA Methodology

The FSA identified 27 potential policy options to mitigate the capability gap the FNA identified. These potential solutions considered changes in doctrine, organization, training, leadership, personnel, and facilities. A screening process winnowed the potential solution options to a smaller set by making a first-order assessment of the potential effects of each option on the recapitalization date and the option's viability given other policy concerns. The solution options that offer potential with minimal associated negative effects or barriers to implementation were then analyzed in greater detail. A more-detailed analysis assessed the remaining options in terms of their potential effect on C-130 fleet life and their potential for closing the capability gap. The most promising options then underwent a net present value (NPV) cost analysis. Integrating the cost and effectiveness analyses provided a means to judge the viability of the potential policy options. SLEPs have high costs that

---

[7] On January 3, 2007, there were 405 MAF C-130E/Hs and 37 C-130Js, for a MAF fleet of 442 aircraft on a TAI basis. Recent budget documents project that the Air Force will acquire an additional 28 MAF C-130Js by the end of FY 2010. The projection is based on Air Force Financial Management and Comptroller, *Committee Staff Procurement Backup Book: FY 2008/2009 Budget Estimates, Aircraft Procurement, Air Force*, Vol. I, Washington, D.C., U.S. Air Force, February 2007.

[8] A new requirement of 335 C-130s is defined in *Mobility Capabilities & Requirements Study 2016*, which was released after the completion of this work. Figure S.1 indicates that the fleet will fall below the 335 requirement in 2017. (DoD, *Mobility Capabilities & Requirements Study 2016*, Washington, D.C., February 26, 2010, Not Available to the General Public.)

Figure S.2
Projected Service Life Remaining for Active and Reserve C-130s

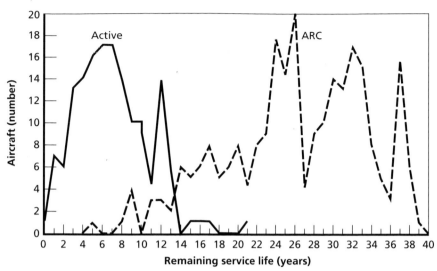

RAND *MG818-S.2*

are on the order of new aircraft acquisitions and were therefore considered material options and analyzed in the UIAFMA. To provide a common metric between aircraft with different lifetime flying profiles, the EBH methodology was developed by the community. EBH provides a common measure of CWB damage for all aircraft. The EBH of each aircraft is tracked individually.

Leverage for Postponing the Need to Recapitalize

All potential solution options identified and analyzed during the FSA fell into one of three broad categories:

- reducing the EBH usage rate of the current C-130 fleet
- increasing the supply of EBH
- meeting the requirement with fewer C-130s.

To get an indication of the leverage each of these three categories of options offered, we made an arbitrary parametric change of 25 percent in each and measured the effects on the required C-130 recapitalization date. Although such changes would not necessarily be possible, the parametric analysis provided useful insights. These charts are intended to provide the reader with a sense of potential delay in the need to recapitalize that could be realized by making fairly significant changes in the three categories. This background should be useful to the reader later in the monograph during the discussion of the specific options we evaluated.

Figure S.3 shows that a 25-percent reduction in the EBH accumulation rate could delay the need to recapitalize by only about two years, to 2015, because so many C-130s are already close to retirement. Figure S.4 shows that an arbitrary 25-percent increase in the amount of EBH available for each aircraft prior to grounding could delay the need to recapitalize by about nine years, to 2022. However, flying

Figure S.3
Reducing Accumulation of Equivalent Baseline Hours by 25 Percent

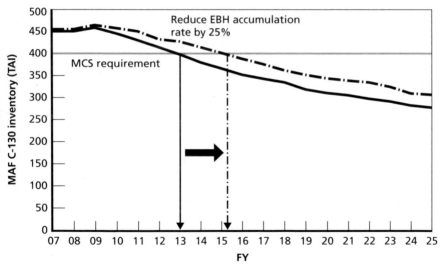

NOTE: Ways to reduce usage rates include the use of simulators, companion trainers, and C-17 substitution.

Figure S.4
Increasing Availability of Equivalent Baseline Hours by 25 Percent

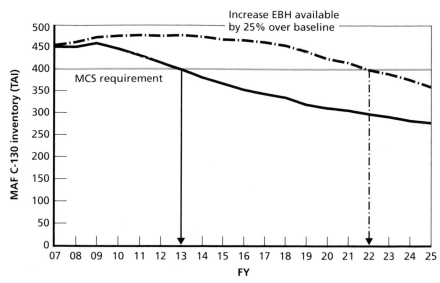

NOTE: Increasing the supply of EBH increases EBH tolerance.
RAND MG818-S.4

beyond 45,000 EBH could entail significant risks of flight failure in
the absence of actions to mitigate structural fatigue damage.[9] Figure
S.5 shows that, if the MCS requirement could be met with 25 percent
fewer C-130s, the recapitalization need could be delayed by about eight
years, to 2021.

Comparison of Potential Solution Options

Tables S.1, S.2, and S.3 show the options considered in each broad
category and their initial screening results. For each option, the potential impact of that option was assessed relative to the other options in
the category (e.g., reducing EBH accumulation). Other implications—

---

[9] The cases shown here give a sense of the leverage broad categories of policy solutions offer
for addressing the capability gap. Increasing the EBH limit on each aircraft by 25 percent to
over 56,000 EBH entails a great deal of risk. We consider this risk to be unacceptably high,
unless significant modifications are undertaken to mitigate CWB structural fatigue issues.

Figure S.5
Meeting Mobility Capabilities Study Requirement with 25-Percent
Fewer C-130s

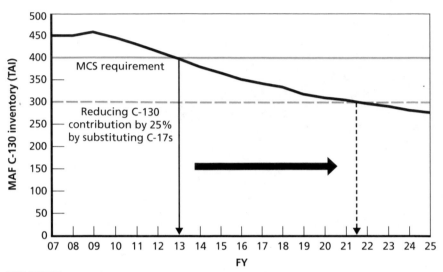

primarily negative implications—are presented. The potential leverage
and impact on the fleet of each of these options must also be considered. Options that had good potential within a category and limited
negative implications were then assessed in a cost-effectiveness analysis
to determine the potential to delay the need to recapitalize the C-130
fleet.

Table S.1 shows potential options for reducing the rate of EBH
accumulation. Although several options could significantly reduce the
rate of EBH accumulation, none could significantly delay the need
to recapitalize a C-130 fleet having so many high-EBH aircraft. The
parametric analysis presented above shows that even a fairly significant
reduction in the rate of EBH accumulation would delay the need to
recapitalize by only one or two years, since many of these aircraft have
only a few years of life remaining at the current operational tempo.[10]

---

[10] For example, a 20-percent reduction in EBH accumulation on an aircraft that has four
years of useful life remaining would extend the life of the aircraft by only one year.

Table S.1
Options for Reducing Equivalent Baseline Hour Accumulation

Options	Estimated Potential Effect	Other Implications
Most-promising		
Increase use of simulators for training	High	None
Increase use of companion trainer aircraft (CTA)	High	None
Shift high-severity-factor operational missions to other aircraft	High	None
Reduce crew qualifications	Moderate	Loss of capacity or flexibility
Reduce high-severity-factor training	Moderate	Loss of capacity or flexibility
Dropped in the screening process		
Rotate aircraft among components	Moderate	May not be viable
Increase experience mix	High	Effects on personnel
Change active-reserve mix	Moderate	Few active units Effect on temporary duty
Add ANG and/or AFRC associate units to active squadrons	Low	Crew ratio cuts needed
Increase squadron size	Very low	Reduced flexibility
Place flight restrictions on specific aircraft	Very low	Reduced flexibility

Key:
Green, *few* or *none*
Yellow, *moderate*
Red, *significant*

Most options that could provide a year or two of delay cost more in NPV terms than procuring new aircraft. All of these are nonviable because of the NPV cost.

The only potential option in this class that resulted in an NPV savings was increased use of simulators. Although potentially delaying recapitalization by only a year or two, this option has a significant NPV savings of about $7 billion—that is, about $200 million per year.

xxiv Intratheater Airlift Functional Solution Analysis

The second broad class of solution options increases the amount of EBH available. Table S.2 presents the potential options evaluated in this category. The parametric analysis showed that this class of option has good potential leverage. There are essentially two ways to do this. The first is a materiel solution (either SLEPing or buying new aircraft), and the second is flying the aircraft beyond 45,000 EBH on the CWB without conducting a SLEP. The SLEP option, a materiel solution, can involve repair, refurbishment, and replacement of structural components having fatigue or corrosion damage. This effectively resets the fatigue damage clock at a lower EBH. An initial assessment indicates that this option may be a cost-effective solution and should be evaluated in a future analysis along with the option of procuring new aircraft. SLEPs and new aircraft acquisitions are materiel solutions. Therefore, detailed analysis of these materiel solutions is left to the AoA.

Flying aircraft beyond 45,000 EBH without a SLEP was assessed as nonviable because of safety concerns. The risk of catastrophic structural failure increases greatly when an aircraft has more than 45,000 EBH on the CWB. Uncertainty about the accumulation of EBH for old aircraft complicates risk assessments. Over 30 to 40 years of usage,

Table S.2
Options for Increasing the Supply of Equivalent Baseline Hours

Options	Estimated Potential Effect	Other Implications
Most promising		
SLEP or repair the aircraft	High	Risks associated with aging aircraft
Buy additional aircraft	High	Additional capability Greater flexibility Reduced risk
Accept greater risk	High	Greater risk of catastrophic failure
Dropped in the screening process		
Develop better diagnostic tools	Moderate	Reduced uncertainty

Key:
Green, *few* or *none*
Yellow, *moderate*
Red, *significant*

fatigue life–monitoring approaches and methods for characterizing mission usage have changed several times. Gaps in the reporting of flight data also introduce uncertainties about the degree of fatigue damage. Moreover, the science of fatigue cracking is not completely understood. Inspections cannot completely compensate, since fatigue cracks are difficult to find and often missed during inspections.[11] At advanced levels of EBH, inspections cannot assure safety for aircraft with widespread fatigue damage. As a result, the amount of life gained from flying beyond the CWB grounding threshold does not appear to justify the significant risk of aircraft losses.

The third set of options evaluated—meeting the requirement with fewer C-130 aircraft—is shown in Table S.3. The parametric analysis showed that reducing the number of C-130s needed to meet the MCS requirement offered good leverage for delaying the fleet recapitalization date. Many of the options shown were dropped in the preliminary screening process because they either had little effect on delaying the recapitalization date or had other negative implications.

Using C-17s in the intratheater role and backfilling the strategic mission with additional Civil Reserve Air Fleet (CRAF) aircraft could potentially reduce the number of C-130s needed to meet the airlift needs identified in the MCS. However, we found this option problematic for several reasons. First, airlift requirements for the "Long War" and potential changes in the way the Army proposes to operate could drive intratheater airlift requirements well beyond those identified in the MCS. Ongoing operations in Iraq and Afghanistan have tied up a large number of C-130s over the last six years. If this level of commitment continues, the Air Force's ability to postpone the need to recapitalize is severely constrained by its need to maintain forces to support the demands of ongoing operations. Further, our analysis of the number of C-130s required to meet the MCS requirement depended on several MCS assumptions that were highly favorable to a C-130/C-17/CRAF

---

[11] For inspection of some fatigue-critical locations on C-130s, Warner Robins Air Logistics Center has assessed the probability of an inspection occurring properly, as specified, as being 0.5. As a result, the Air Force requires some critical inspections to be performed twice, with independent inspectors and engineering oversight, to raise the probability of a proper inspection to 0.75. Even with this heightened probability of success, cracks can be missed.

Table S.3
Options for Meeting the Requirement with Fewer C-130s

Options	Estimated Potential Effect	Other Implications
Most promising		
Shift some of strategic lift burden to CRAF and some C-17s to theater lift	High	None[a]
Shift more Air Education and Training Command (AETC) aircraft during peak demand	Low	None
Dropped in the screening process		
Shift some of theater lift burden to surface lift	High	Solution options may not be robust
Fly strategic airlift to FOLs	Moderate	Solution options may not be robust
Change theater routes	Low	Solution options may not be robust
Increase maximum number of aircraft on the ground (more civil engineering)	Low	Solution options may not be robust
Increase crew ratio	Low	Solution options may not be robust
Use Joint Precision Air Drop System	Low	Longer load times More training and qualification
Increase Army days of supply	Low	May increase need for tails
Pool joint airlift	Low	
Reduce number of aircraft subjected to a change in operational control (CHOPed)	None	None
Improve in-transit visibility	None	None

Key:
Green, *few* or *none*
Yellow, *moderate*
Red, *significant*

[a] The rating for this option reflects our initial screening. Further analysis indicated that this option is unworkable, principally because meeting the MCS requirement with fewer C-130s could leave the Air Force with inadequate force structure for sustained operations (i.e., the Long War requirement).

swap. As a result of the potential increased need for intratheater airlift beyond the scope of the MCS and the potential fragility of the option because of the favorable MCS assumptions, we judged the C-17 and CRAF substitution option not to be viable.

Conclusion

Table S.4 summarizes the assessment of options that underwent a detailed cost-effectiveness analysis. We found no viable nonmateriel solution or combination of nonmateriel solutions that could delay the need to recapitalize the fleet by more than a few years. Since no viable nonmateriel solution was identified in the FSA, an AoA should be

Table S.4
Summary of Results

FSA Option	Delays Need for Recapitalization by (years)	NPV	Other Implications
Meet MCS requirement with fewer C-130s			
Shift some C-17s to theater role; backfill with CRAF in strategic airlift			Long War dominates: Not viable
Shift more AETC aircraft during peak demand	~1–2		
Reduce EBH usage rate			
Shift more training to simulators	1–2	A savings of $7 billion	
Use CTA	1	A cost of $6 billion	
Shift some contingency missions to other mission design series	<1	A cost of $2 billion	
Increase EBH supply			
Fly aircraft beyond 45,000 EBH (fly to 56,000 EBH)	~9	Uncertain	Unacceptably high risk
SLEP	~20	TBD	

undertaken to evaluate potential materiel solutions, including SLEPs and new aircraft buys.[12]

---

[12] See Kennedy et al., 2010. This FSA has deferred in-depth analysis of SLEPs and new aircraft buys to the UIAFMA, which is more appropriate for these materiel solutions.

Acknowledgments

The authors are grateful to many individuals in the U.S. Air Force and the U.S. government. Maj Gen Thomas Kane, Director of Strategic Plans, Requirements, and Programs, Headquarters AMC, was the sponsor of this work and provided support and guidance during the study. Our action officers in AMC/A5QH—Wayne Armstrong, Lt Col Eugene Capone, Craig Lundy, and Michael Houston—were an enormous asset throughout this study. Wayne, Gene, Craig, and Mike provided critical feedback throughout the work. In addition, they were always able to guide us to the sources we needed for data and information. In addition, they often coordinated meetings. They also took care of a variety of tasks, including setting up meetings and briefings, distributing charts, and a variety of administrative tasks. They always went the extra mile to ensure that everything was done correctly and functioned well.

We would also like to thank other folks at AMC who provided data, information, and critical review of our work. At AMC/A9, we would like to thank Dave Merrill, Randy Johnson, and Lt Chris Jones. We are grateful to Craig Vara at AMC/A3TF for helping us better understand the C-130 training program and Phil Widincamp at AMC/A3TR for proving the base model for the flying-hour program. Maj Jessie Vickers from AMC/A4MJ provided critical information about inspection and repair. At AMC/A8PF, we would like to thank Maj Michael Honma for providing vital information throughout the study. Lt Col John Raymond and Maj Paul Cook at AMC/A3OO were critical to our analysis of the ongoing operations. During this study, we gained an enormous amount of respect for the mobility professionals at AMC.

We received an enormous amount of help throughout this study from individuals on the Air Staff. Harry Disbrow, AF/A5R, provided support and guidance throughout this study. We are grateful to Mr. Disbrow for sharing his thoughts and for his critical review of our work. Lt Col Frank Altieri at AF/A5RM and Lt Col Jeffrey Brown at A5XC/GM provided information and feedback throughout the study. We are fortunate that Frank and Jefe were involved in this study and we thank them for their efforts.

We received a great deal of help on C-130 aging and structural issues from 330th Agile Combat Support. Peter Christiansen, Jay Fiebig, and Marian Fraley spent a great deal of time working with us and sharing the wealth of knowledge that they have accumulated on the health of the C-130 fleet. The 330 ASC also provided a great deal of data that were required for this study. We very much appreciate the help they provided during the study.

We would like to express our gratitude to the Director of the Fleet Viability Board, Francis Crowley. Fran provided some key feedback and comments during the study and worked to ensure that we received the information and data we required. We are grateful for his efforts.

During the course of this study, we convened two integrated product team meetings to explore and evaluate the potential solution options. These meetings were very helpful during the study and we are grateful to the team members: Steve Baca, Paul Caster, Tim Gannon, Col Cal Lude, David Hammond, Heather Hendrickson, John Krieger, George Pruitt, Col Robert Quackenbush, Tom Simoes, and Laura Williams.

We would like to thank several of our RAND colleagues. Jean Gebman shared insights from his long research on aging aircraft issues. Lt Cols Larry Gatti and John Wood provided an enormous amount of help during the course of our work while USAF Fellows at RAND. John Stillion was instrumental in setting up this study and provided critical feedback and assistance. John led the preceding study, which culminated in the intratheater FAA and FNA, but was needed for other Project AIR FORCE priorities and was unable to lead this work.

We would like to thank Phyllis Gilmore for her great editing work on an earlier draft. We would also like to thank the technical reviewers, David Shlapak, Robert Ernst, and Mahyar Amouzegar. The monograph is much stronger as a result of their thoughtful reviews. Finally, John Ausink provided helpful comments and suggestions on a portion of the draft.

Abbreviations

AFB	Air Force base
AETC	Air Education and Training Command
AFI	Air Force instruction
AFRC	Air Force Reserve Command
AFSOC	Air Force Special Operations Command
AFTOC	Air Force Total Ownership Cost
AIRCAT	Automated Inspection, Repair, Corrosion and Aircraft Tracking
AMC	Air Mobility Command
AMC/A3	Air Mobility Command, Director of Operations
AMC/A3T	Air Mobility Command, Aircrew Operations and Training Division
AMC/A3XP	Air Mobility Command, Regional Plans Branch
AMC/A4MJ	Air Mobility Command
ANG	Air National Guard
AoA	analysis of alternatives
ARC	Air Reserve Component
ASIP	Aircraft Structural Integrity Program

AUPC	average unit procurement cost
BCT	brigade combat team
BRAC	Base Realignment and Closure
CA	combat aircraft
CASCOM	Combined Arms Support Command
CAStLE	Center for Aircraft Structural Life Extension
CBA	capabilities-based assessment
CBRNE	chemical, biological, radiological, nuclear, and explosive
CHOPed	change in operational control
CJCS	Chairman of the Joint Chiefs of Staff
CJCSI	Chairman of the Joint Chiefs of Staff instruction
CLS	contractor logistics support
CONUS	continental United States
CORE	cost-oriented resource estimating
CRAF	Civil Reserve Air Fleet
CWB	center wing box
CY	calendar year
del	delivery
DoD	Department of Defense
DoS	days of supply
DOTMLPF	doctrine, organization, training, materiel, leadership and education, personnel, and facilities
EBH	equivalent baseline hour(s)

ECP	engineering change proposal
FAA	functional area analysis
FMA	fleet mix analysis
FNA	functional needs analysis
FOL	forward operating location
FSA	functional solution analysis
FY	fiscal year
GAO	Government Accountability Office [formerly General Accounting Office]
GDSS	Global Decision Support System
GWOT	global war of terrorism
IPT	integrated product team
IRT	independent review team
JA/ATT	oint airborne and air transportability training
JCIDS	Joint Capabilities Integration and Development System
JP	joint publication
JPADS	Joint Precision Airdrop System
MAF	mobility air forces
MCS	Mobility Capabilities Study
MDS	mission design series
MED	multiple element damage
MIL-STD	military standard
MILCON	military construction

MOB	main operating base
MSD	multiple site damage
NDI	nondestructive inspection
NDS	National Defense Strategy
NPV	net present value
NSS	National Security Strategy
O&S	operations and support
ODS	Operation Desert Storm
OEF	Operation Enduring Freedom
OIF	Operation Iraqi Freedom
OSD	Office of the Secretary of Defense
PAA	primary aircraft authorized
PA&E	Program Analysis and Evaluation
PDM	programmed depot maintenance
PMAI	primary mission aircraft inventory
PME	Professional Military Education
POD	probability of detection
POI	probability of inspection
PQP	prior-qualified pilots
R&D	research and development
SAAM	special assignment airlift mission
SAR	selected acquisition report
SHM	structural health monitoring
SLEP	service-life extension program

SOF	special operations forces
TAI	total aircraft inventory
TBD	to be determined
TCTO	time compliance technical order
TDY	temporary duty
TRADOC	U.S. Army Training and Doctrine Command
TS/MC	time-sensitive, mission-critical
TWCF	Transportation Capital Working Fund
UIAFMA	U.S. Air Force Intratheater Airlift Fleet Mix Analysis
UPT	undergraduate pilot training
USAF	U.S. Air Force
USAFE	U.S. Air Forces in Europe
WMD	weapons of mass destruction
WME	weapons of mass effect
WR-ALC	Warner Robins Air Logistics Center

CHAPTER ONE

Introduction and Background

This functional solution analysis (FSA) for U.S. Air Force (USAF) intratheater airlift is the third in a series of three documents that together constitute a capabilities-based assessment (CBA) required as part of the Joint Capabilities Integration and Development System (JCIDS).[1] These three documents of the CBA are part of a top-down process within that system for identifying and assessing capability needs.

The first document in the series, the functional area analysis (FAA),[2] identified the operational tasks, conditions, and standards needed to achieve military objectives—in this case, certain intratheater airlift missions. The second, the functional needs analysis (FNA),[3] assessed the ability of current assets to deliver the capabilities identified in the FAA. The third document in the series, the FSA, is an operationally based assessment of current capabilities to determine whether a nonmateriel solution could be used to close any capability gaps identified in the FNA.

---

[1] Chairman of the Joint Chiefs of Staff Instruction (CJCSI) 3170.01E, *Joint Capabilities Integration and Development System (JCIDS)*, Washington, D.C., May 11, 2005. establishes the policies and procedures of the JCIDS process.

[2] Documented in David T. Orletsky, Anthony D. Rosello, and John Stillion, *Intratheater Airlift Functional Area Analysis (FAA)*, Santa Monica, Calif.: RAND Corporation, MG-685-AF, 2011.

[3] Documented in John Stillion, David T. Orletsky, and Anthony D. Rosello, *Intratheater Airlift Functional Needs Analysis (FNA)*, Santa Monica, Calif.: RAND Corporation, MG-822-AF, 2011.

Thus, the broad objective of the FAA, FNA, and FSA is to determine whether a materiel solution is required to address specific shortfalls in military capabilities or whether modifications to other aspects of the system could resolve the shortfall. If the FSA determines that no nonmateriel solution can address the shortfall, an analysis of alternatives (AoA) is normally undertaken to evaluate the cost-effectiveness of various materiel solutions.

This FSA is an operationally based assessment of approaches to addressing the gap identified in the FNA. If no viable nonmateriel solutions are identified in the FSA, an AoA is normallly undertaken to evaluate potential materiel solutions.[4]

Capabilities-Based Assessment Objectives

This CBA was initiated to analyze a potential deficiency in intratheater delivery capability, in response to concerns that demands from the ongoing global war on terrorism (GWOT) and new operational concepts the U.S. Army has been considering may result in a shortfall in USAF capabilities to deliver personnel and equipment to increasingly numerous and dispersed theater operating locations.

This CBA focuses on the intratheater cargo and personnel movement mission, which is driven primarily by the joint land force requirement to move personnel, equipment, and supplies throughout the battlespace. The Army has already completed an FAA, FNA, and FSA on Army Fixed Wing Aviation that identified several shortfalls requiring a materiel solution.[5] Quadrennial Defense Review 2005 requires a joint

---

[4] For this study, the relevant results are being presented in Michael Kennedy, David T. Orletsky, Anthony D. Rosello, Sean Bednarz, Katherine Comanor, Paul Dreyer, Chris Fitzmartin, Ken Munson, William Stanley, and Fred Timson, *USAF Intratheater Airlift Fleet Mix Analysis* [UIAFMA], Santa Monica, Calif.: RAND Corporation, 2010, Not Available to the General Public. Because service-life extension programs (SLEPs) and new aircraft buys are considered materiel solutions, the UIAFMA, rather than this FSA, covers the relevant analyses.

[5] U.S. Army Aviation Center, Futures Development Division, Directorate of Combat Developments, *Army Fixed Wing Aviation Functional Area Analysis Report*, Fort Rucker, Ala., June 3, 2003a; U.S. Army Aviation Center, Futures Development Division, Director-

program office to implement the acquisition of any aircraft procured as a result of the Army studies.[6] Near the end of calendar year (CY) 2005, the USAF asked RAND to conduct the USAF intratheater delivery F-series studies. In February 2006, the Chiefs of Staff of the USAF and Army signed a memorandum of understanding that directed the services to develop a joint memorandum of agreement within 90 days to articulate the path forward for each of the services toward developing complementary capabilities with respect to light cargo aircraft.[7] Figure 1.1 shows the timing of the various studies. This figure shows that the Army studies were completed in 2003 through 2005, with the addendum to the Army AoA being completed in 2007. The USAF F-series studies were conducted in 2006 and 2007. The UIAFMA was conducted after the FSA to evaluate potential materiel solutions. The UIAFMA was conducted using a cost-effectiveness analysis and could be considered the USAF AoA.[8] Next we discuss the FAA and FNA.

Functional Area Analysis

The FAA identified three broad operational mission areas for intratheater airlift:[9]

---

ate of Combat Developments, *Army Fixed Wing Aviation Functional Needs Analysis Report*, Fort Rucker, Ala., June 23, 2003b; U.S. Army Aviation Center, Futures Development Division, Directorate of Combat Developments, *Army Fixed Wing Aviation Functional Solution Analysis Report*, Fort Rucker, Ala., June 8, 2004; U.S. Army Training and Doctrine Command (TRADOC) Analysis Center, *Future Cargo Aircraft (FCA) Analysis of Alternatives (AoA)*, Fort Leavenworth, Kan., TRAC-TR0-5-18, July 18, 2005b, Not Available to the General Public. TRADOC Analysis Center, *Joint Cargo Aircraft/Future Cargo Aircraft Analysis of Alternatives*, final results scripted brief, Fort Leavenworth, Kan., TRAC-F-TR-07-027, March 2005a.

[6] Department of Defense (DoD), *Quadrennial Defense Review Report*, Washington, D.C., February 6, 2006.

[7] U.S. Army and U.S. Air Force, "Way Ahead for Convergence of Complementary Capabilities," memorandum of understanding, February 2006.

[8] Kennedy et al., 2010.

[9] Meeting at Air Mobility Command (AMC), December 8, 2005, and subsequent discussions with USAF personnel.

Figure 1.1
Timing of Army and USAF Studies

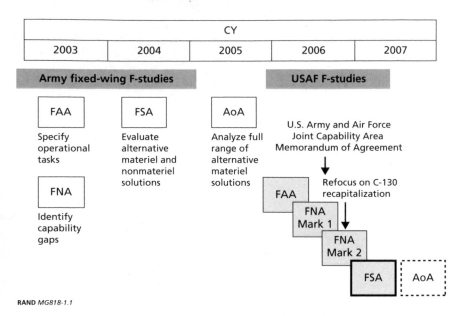

RAND *MG818-1.1*

- routine sustainment—the steady-state delivery of required supplies and personnel to units
- time-sensitive, mission-critical resupply—the delivery of supplies and personnel on short notice in support of deployed Army units, outside the steady-state demands
- maneuver—the transport of combat teams around the battlefield using the intratheater airlift system.

These three operational mission areas have different characteristics and impose different requirements on the intratheater airlift system.

The FAA used multiple sources for input and guidance, including the National Security Strategy, the National Defense Strategy, the Joint Operations Concepts Family, the U.S. Air Force Flight Plan 2004, the Global Mobility Concept of Operations, and Army Vision. We also consulted the Universal Joint Task List and the Air Force Master Capa-

Table 1.1
Tasks/Mission Areas Applicable to This CBA

Task	Routine Sustainment	Time-Sensitive, Mission-Critical Resupply	Small-Unit Maneuver
Transport supplies and equipment to points of need	X	X	X
Conduct retrograde of supplies and equipment	X	X	X
Transport (deployment, redeployment, and retrograde) of forces and accompanying supplies to point of need			X
Conduct recovery of personnel and supplies (including evacuation of hostages, evacuees, enemy personnel, and high-value items)			X
Transport replacement/ augmentation personnel	X	X	X
Evacuate casualties	X	X	X

bility Library.[10] Table 1.1—adapted from the FAA—presents the tasks derived from these sources and identifies their applicability to each of the three mission areas in this CBA.

Although the guidance documents do not specify a set of conditions under which these tasks must be accomplished, they do discuss attributes and conditions. Some of these attributes and conditions occur in multiple guidance documents. We selected from among these.

Among the conditions deemed important to consider in this CBA were such concerns as adverse weather; multiple, simultaneous, distributed decentralized battles and campaigns; degraded environments; and infrastructure issues. Desirable attributes and conditions include a small logistical footprint; speed, accuracy, and efficiency; and basing flexibility.[11]

---

[10] A complete discussion of these guidance documents can be found in Orletsky et al., 2011, pp. 5–25 and 31–35.

[11] See Chapter Two of Stillion, Orletsky, and Rosello, 2011, for specifics.

The guidance documents also specify standards for evaluating potential capability gaps, such as the ability to provide materiel support for current and planned operations in optimal cycle times.

Functional Needs Analysis

The FNA identified and analyzed two potential capabilities gaps. The first involves the need to maintain enough C-130s to meet the requirement identified in the Mobility Capabilities Study (MCS).[12] The MCS set the minimum number of USAF mobility air forces (MAF) C-130s at 395 (total aircraft inventory [TAI]). The second is to provide responsive intratheater resupply in support of the U.S. Army. The FNA looked at providing both routine sustainment and time-sensitive, mission-critical resupply of a sizable multibrigade combat team (BCT) ground force.

A significant and growing portion of the C-130 fleet is either restricted or grounded because of fatigue-related cracking of key structural components of the center wing box (CWB). At the beginning of CY 2007, a total of 442 C-130 aircraft were assigned to MAF,[13] including all aircraft assigned to active Air Force, Air Force Reserve Command (AFRC), and Air National Guard (ANG) units whose primary missions are either airlift or training airlift crews.[14] Even under optimistic assumptions, the FNA found that, if the policies of imposing

---

[12] The MCS was conducted by the Director of Program Analysis and Evaluation (PA&E) and the Chairman of the Joint Chiefs of Staff (CJCS) in collaboration with the Office of the Secretary of Defense (OSD), the Joint Staff, services, and combatant commands (DoD and the Joint Chiefs of Staff, *Mobility Capabilities Study*, Washington, D.C., December 19, 2005, Not Available to the General Public). Throughout this document, the study is subsequently referred to simply as *the MCS*.

[13] Throughout this document, we will refer to the requirement and fleet in terms of MAF C-130s. As of January 3, 2007, the MAF C-130 fleet consisted of 405 C-130E/Hs and 37 C-130Js. Additional C-130J aircraft are scheduled for delivery: seven additional aircraft by the end of fiscal year (FY) 2007, eight in FY 2008, seven in FY 2009, and six in FY 2010.

[14] The 442 MAF aircraft do not include the LC-130s and the WC-130s because these special-mission aircraft are specially configured and fly specific nonmobility missions. Although these aircraft can and do fly AMC missions, they may not always be available.

flight restrictions and grounding aircraft remain in place and nothing else is done, the number of unrestricted C-130s available to the USAF is likely to fall below the minimum threshold of 395 by 2013 because of fatigue-related aircraft groundings.[15]

In addition, the FNA examined some of the factors that might cause the demand for intratheater airlift capability to increase beyond the MCS's full military mobilization requirement of 395 C-130s. These include increased reliance on air delivery and proposed U.S. Army operations concepts that call for highly dispersed operations, which are resupplied entirely by air.[16] Although having large multi-BCT forces operate without a ground line of communication is not the current Army concept for future operations, the trend is toward more-dispersed operations of ground forces. Future ground forces will rely on increased aerial distribution.[17]

The FNA found that it would be extremely challenging to routinely supply 100 percent of the daily sustainment of a moderate-size ground combat force using the existing intratheater airlift system. Existing intratheater airlift assets can provide a robust, responsive time-sensitive, mission-critical resupply system with a reasonably small commitment of resources. In addition, allocating additional resources to this mission beyond the levels we chose results in rapidly diminishing returns in terms of reduced time in transit. This, combined with the fact that time in transit accounts for only part of the total time between request and delivery, suggests that investments in improving logistics management processes and procedures may be a more fruitful

---

Further, the special equipment may limit the amount and type of cargo they can carry. As a result, we did not include these aircraft in the MAF aircraft in this analysis.

[15] This number assumes that current grounding policies remain in place and that each aircraft accumulates fatigue damage at the same pace as in the past. These projections are based on the C-130 System Program Office's Automated Inspection, Repair, Corrosion and Aircraft Tracking (AIRCAT) database, C-130 AIRCAT Center Wing Equivalent Baseline Hours (EBH) Report, spreadsheet, January 3, 2007; see also Chapter Three.

[16] By contrast, the MCS assumed that only a fraction of the daily sustainment of ground forces would be provided by air.

[17] See U.S. Army Aviation Center, 2003b, p. 16-17.

means of realizing substantial reductions in the total time-sensitive, mission-critical resupply performance.

Because routine resupply is not a requirement and because time-sensitive, mission-critical resupply takes relatively few assets, the FNA determined that the FSA should focus on ensuring that the intratheater airlift fleet continues to meet the 395 C-130 requirement identified in the MCS. This requirement is at issue because of the large number of aircraft that are expected to undergo flight restrictions and groundings during the next two decades. Figure 1.2 plots each aircraft's unique annual flying rate and EBH accumulation rate, illustrating the projected decline in the MAF C-130 inventory as aircraft reach the grounding limit.[18] The number of C-130s is projected to fall below the MCS requirement of 395 in 2013.

Organization of This Document

Chapter Two presents a detailed evaluation of the health of C-130 fleet and the EBH methodology. Chapter Three describes the FSA analytical methodology and the three broad classes of solution options. Chapter Three also presents some parametric evaluation of these broad classes of solution options to provide a sense of where the leverage is to close the capability gap. Chapters Four, Five, Six, and Seven present the detailed analysis of the solution options. Chapter Eight describes the cost analysis and presents the analysis of several policy options. Chapter Nine presents the conclusions of the FSA. The appendix provides additional information on structural issues.

---

[18] On January 3, 2007, there were 405 MAF C-130E/Hs and 37 C-130Js for a MAF fleet of 442 aircraft (TAI). Recent budget documents project that the Air Force will acquire an additional 28 MAF C-130Js by the end of FY 2010. The projection is based on Air Force Financial Management and Comptroller, *Committee Staff Procurement Backup Book: FY 2008/2009 Budget Estimates, Aircraft Procurement, Air Force*, Vol. I, Washington, D.C., U.S. Air Force, February 2007.

Figure 1.2
Decline in the MAF C-130 Inventory as Aircraft Reach the Grounding Limit

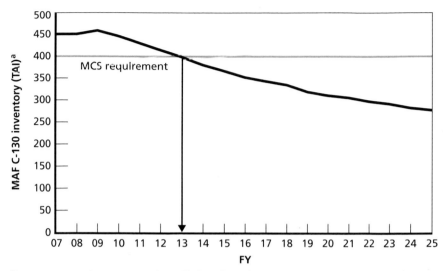

ªInventory numbers assume that all aircraft undergo TCTO 1908 inspection and are able to fly 45,000 EBH.

RAND *MG818-1.2*

CHAPTER TWO

Intratheater Airlift Fleet Condition and Status

C-130s perform many diverse missions for USAF and other operators. This FSA focuses on the C-130s used in the air mobility mission for intratheater transport of personnel and materiel. Corrosion and fatigue damage in the aging air mobility fleet are contributing to flight restrictions, groundings, and retirements, reducing the number of C-130s available to perform air mobility missions, and prompting recapitalization considerations.

To provide a context for subsequent analysis, this chapter describes the constitution of the C-130 air mobility fleet, the distribution of aircraft by operator, the age and flying status of the various models, the pedigree of key structural components, and their structural health. The chapter also briefly discusses functional systems that may also require maintenance, upgrading, or replacement as C-130s get older. The appendix presents additional detail on the current health of the C-130 fleet, including the evolution of key structural components, issues affecting service life, and the uncertainty of health of the fleet.

Fleet Composition and Status

On January 3, 2007, the MAF fleet included 405 C-130E/Hs (TAI) and 37 C-130Js, for a total fleet of 442 aircraft.[1] Figure 2.1 is a snap-

---

[1] The spreadsheets drawn from the AIRCAT database used in this chapter are distinct from the report discussed at length in Chapter Three. For the sake of clarity, for general CWB data, we cite AIRCAT and include the specific date. We discuss other aging systems in Chapter Two.

Figure 2.1
C-130E/H Center Wing Box Structural Problems Are Reducing the Number
of Aircraft Available for Air Mobility Missions

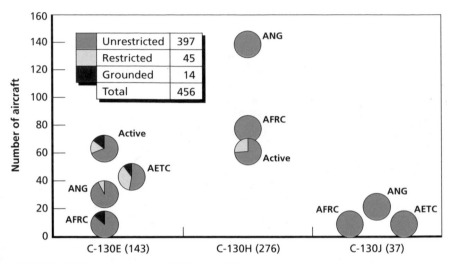

SOURCES: CWB sheet, January 2007; AIRCAT 2007.
NOTES: Includes AMC, USAFE, Pacific Air Forces, AETC, ANG, and AFRC aircraft;
excludes aircraft retired or converted to ground trainers. The center of each circle is
positioned vertically to reflect the number of aircraft, as indicated on the y-axis. Data
are as of January 3, 2007.
RAND *MG818-2.1*

shot of the size, constitution, and flying status of the air mobility fleet
at the beginning of 2007. At that time, 397 C-130Es, Hs, and Js were
on unrestricted flying status, and 45 Es and Hs were operating under
flight restrictions because of the risks accumulated fatigue damage
poses. The figure shows an additional 14 Es that had been grounded
because of fatigue damage but not yet retired. Groundings and flight
restrictions have fallen most heavily on the active and training components, which operate more of the older E- and H1-model aircraft.
Reserve components operate more of the newer H- and J-model aircraft and have been less affected by restrictions and groundings. The
operating restrictions and the distribution of aircraft across components influence the flexibility with which USAF can satisfy operational
taskings.

Retirements driven by the fatigue and corrosion problems of C-130Es and, to a lesser extent, C-130Hs are reducing the size of the air mobility fleet and are prompting examinations of the need to recapitalize intratheater transport. The remainder of this chapter will characterize the aging of C-130Es and Hs and assess the implications of that aging.

Because procurement of the current C-130 air mobility fleet dates back to the 1960s, many C-130s, and especially C-130Es, are very old in chronological terms (33 to 44 years) and in usage terms (most have 20,000 to 30,000 flight hours); see Figure 2.2. Aircraft with these demographics are at particular risk of corrosion and metal fatigue.[2] The

Figure 2.2
The C-130E/H Air Mobility Fleet Has a Significant Population of Aging Aircraft

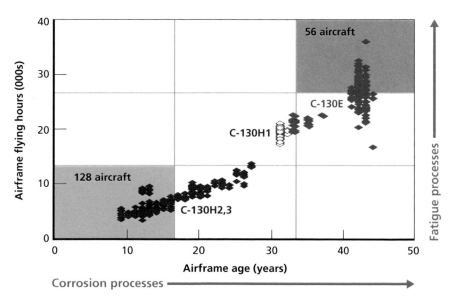

SOURCES: CWB sheet, January 2007; Lockheed Martin Aeronautics Company (LMAC), Fleet Viability Board briefings, December 6, 2006.
RAND MG818-2.2

_____

[2] Although much of the attention on C-130 aging has focused on structural problems, other functional systems also require attention as aircraft age. We will discuss other aging systems later in this chapter.

aircraft in the upper right portion of this chart are most at risk. Significant corrosion and fatigue problems are likely for aircraft in this group. A group of 47 C-130H1s was produced immediately after conclusion of C-130E production and is more than 30 years old. The remaining H2s and H3s range in age from 9 to 27 years.[3]

Figure 2.3 illustrates some manifestations of the aircraft aging just described. Because of CWB fatigue damage, USAF has imposed flight restrictions on C-130Es and H1s. Some C-130Es have been grounded because of fatigue damage, and 84 C-130Es have been retired during the past four years because of fatigue and corrosion damage. The operating restrictions and retirements have significantly reduced the pool of aircraft available to fulfill air mobility taskings.[4] With the aging population of airplanes depicted in Figures 2.2 and 2.3, these trends are expected to continue.

Aging of airframe structures, engines, and systems falls into several categories:

- metal fatigue
- corrosion
- wear
- material breakdown
- obsolescence
- diminishing manufacturing sources.

Any one of these problems or combination of them may render an aircraft unsafe for flight or uneconomical to repair.[5]

---

[3] Lockheed continues to deliver C-130Js to the Air Force, with the oldest C-130J assigned to air mobility forces being about ten years old (AIRCAT C-130J inventory data downloaded January 2007).

[4] Flight restrictions limit usage to approximately 60 percent of design loads. These restrictions include limits on weights, airspeeds, altitudes, maneuver load factors, and abrupt maneuvers. These restrictions effectively limit planes to certain training missions. See Marian Fraley, "C-130 Center Wing Status," Robins Air Force Base (AFB), Ga.: WR-ALC, February 16, 2005.

[5] The cyclic loading of aircraft structures that takes place on every flight leads to fatigue cracking that can ultimately render an airplane unsafe to fly or uneconomical to repair. Over time, environmental effects can lead to corrosion that can damage aircraft structures

Figure 2.3
Structural Problems Have Led to Flight Restrictions, Groundings, and Retirements

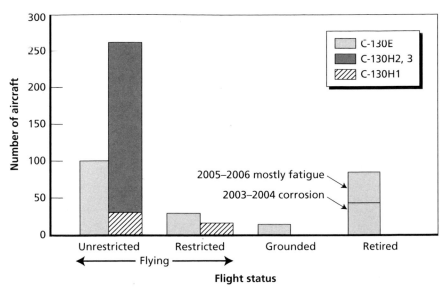

SOURCES: CWB sheet, January 2007; LMAC, 2006.
RAND *MG818-2.3*

Operators face the threat of "modification creep" as successive aging issues present themselves, each requiring additional investments to keep aircraft flying. Uncertainties about the structural health of older C-130s complicate determinations about whether to continue investing in older planes or to retire them. Old, unrepresentative fatigue tests, dated knowledge about flight loads, and uncertainty about past

such that aircraft become unsafe to fly or uneconomical to repair. Fatigue and corrosion processes can be particularly insidious, since damage in inaccessible areas may be hidden from view, becoming apparent only when an aircraft is opened up for maintenance. Parts can also simply wear out with repeated usage. Materials can break down over time because of exposure to heat, gases, fuels, or other factors. Some systems become obsolescent as new technologies are introduced. When aircraft are retained for 40 years or more, vendors go out of business, and it can become harder and harder to find qualified sources for replacement parts. All these factors pose risks to the continuing availability of aging fleets, such as C-130E/Hs.

and current aircraft usage make it difficult to assess the state of fatigue damage of each aircraft and associated risks.[6]

C-130 Structural Issues and the Equivalent Baseline Hours Metric

Figure 2.4 illustrates the key structural components of C-130s. Fatigue cracking in the CWB has led to additional inspection and repair actions, as well as flight restrictions, groundings, and retirements. The outer wings attach to the CWB via so-called *rainbow fittings*. The fuselage consists of forward, center, and aft elements. The empennage consists of the horizontal and vertical stabilizer.

The Air Force and the C-130 contractor, the Lockheed Martin Aeronautics Company (LMAC), have adopted a metric called EBH to express the accumulation of fatigue damage in critical areas of the C-130 structure. The rate of fatigue damage accumulation is referenced to typical C-130E usage in 1981, which was generally more benign than current usage.[7] Airplanes today typically experience higher loads, notionally illustrated in the upper panel of Figure 2.5 by the black line. These higher loads result in more fatigue damage (cracking) per flying hour than was the case with the baseline usage. As illustrated in the lower panel of Figure 2.5, fatigue cracks reach a critical length in fewer flight hours. Alternatively, for the same number of flight hours, airplanes today tend to accumulate more fatigue damage, that is, experience more fatigue cracking.

Figure 2.6, drawn from actual C-130 flight experience, illustrates that, on some particularly damaging high-speed, low-level missions, CWBs accumulate fatigue damage at 2 to 6 times the rate of the 1981

---

[6]   Warner Robins Air Logistics Center (WR-ALC), *Aircraft Structural Integrity Program Master Plan for C-130 Aircraft*, Robins AFB, Ga.: U.S. Air Force Materiel Command, December 1995.

[7]   J. A. Lindenbaum, *Equivalent Baseline Hours Methodology of U.S. Air Force C-130E-H Individual Aircraft Tracking Program (IATP), Automated Inspection, Repair, Corrosion and Aircraft Tracking (AIRCAT)*, Atlanta, Ga.: Lockheed Martin Aeronautics Company, LG07ER0221, February 12, 2007.

Figure 2.4
Principal C-130 Structural Components

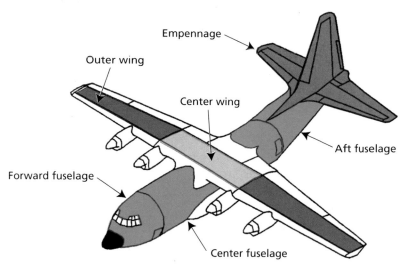

SOURCE: Adapted from illustration in LMAC, 2006.
NOTE: CWB fatigue cracking has led to recent imposition of flight
restrictions on groundings.
RAND *MG818-2.4*

baseline. The so-called *severity factor* relates the actual rate of fatigue damage to that incurred with 1981 baseline usage:

$$SF_A = \frac{Baseline\ service\ life}{Mission\ A\ service\ life}.$$

The product of the severity factor, specific to a particular structural component and location, and actual flight hours, yields EBH:

$$EBH = SF\left(Actual\ flying\ hours\right),$$

where EBH is the number of hours a C-130E in 1981 would have had to fly to incur the same amount of fatigue damage as airplanes flying typical missions today.[8] The most life-limiting component of the C-130

---

[8]   S. F. Ramey and J. C. Diederich, "Operational Usage Evaluation and Service Life Assessment," presented at the 2006 Hercules Operators Conference, Atlanta, Ga., October 2006;

Figure 2.5
Fatigue Damage Accumulation Expressed in Terms
of Equivalent Baseline Hours

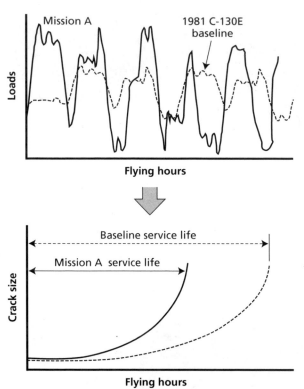

SOURCES: U.S. Air Force, 2007; WR-ALC, 1995; LMAC, 2006;
Air Mobility Support Engineering, "C-130 Center Wing:
Service Life Issues," "C-130 Center Wing: History &
Overview," and "C-130 Center Wing: Replacement
Program," *Lockheed Martin Service News*, Vol. 30, No. 2,
2005, pp. 3–8.
RAND *MG818-2.5*

is the CWB. Recent average C-130E and H squadron severity factors
for fatigue-critical locations in the CWB have ranged from 1.6 to 2.9,
aggregated across the mix of missions flown. The average severity factor
is about 2.15 for all air mobility C-130Es and Hs.[9] The severity factor

---

Lindenbaum, 2007.

[9] AIRCAT C-130J data, January 2007.

Figure 2.6
Severity Factor for Different Missions

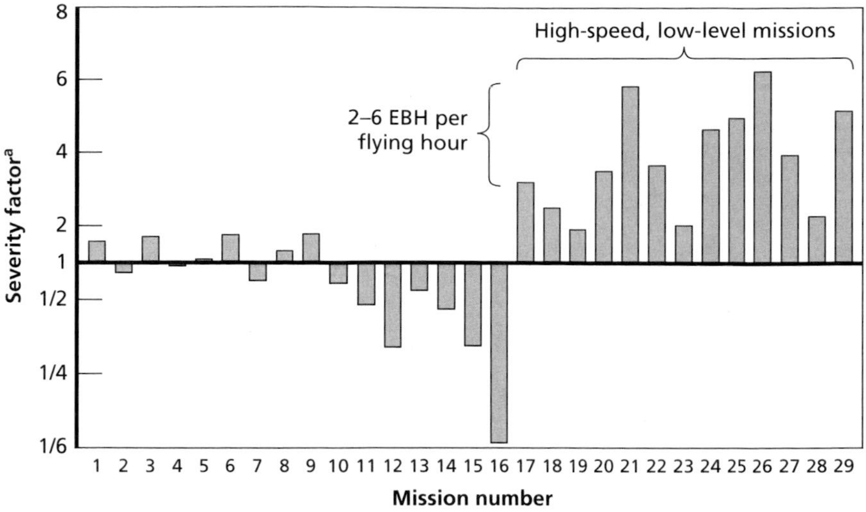

aRelative to USAF C-130E baseline.

for the outer wings is on the order of 1.86.[10] This means that different fatigue-critical locations on an airframe (such as distinct locations in the CWB and the outer wing) can have different EBH accumulations while having the same number of flight hours.

Figure 2.7 groups aircraft in 5,000-EBH cohorts to show the distribution of EBH across the C-130 fleet (5,000 to 10,000-EBH; 10,000 to 15,000 EBH; etc.). A comparatively large number of C-130Es and H1s have passed or are nearing the 38,000-EBH flight restriction threshold or the 45,000-EBH grounding limit.[11] As of this writing, aircraft in the area beyond the dashed line at 45,000 EBH had been grounded but not yet officially retired. The next figure will illustrate the progression of the types of fatigue damage just described relative to the various C-130 EBH cohort groups.

---

[10] Ramey and Diederich, 2006.

[11] WR-ALC, "C-130 Center Wing Status," briefing, February 9, 2005.

Figure 2.7
Distribution of Center Wing Box Equivalent Baseline Hours for C-130E/H

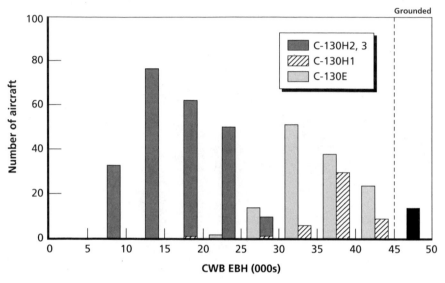

SOURCES: CWB sheet, January 2007; LMAC, 2006.
NOTES: Depicts the number of aircraft that have been grounded but whose status
has not yet been changed to retired. For a more-complete picture of retired aircraft,
see Figure 2.3.
RAND *MG818-2.7*

To illustrate which aircraft groups face the most serious fatigue cracking, Figure 2.8 overlays the typical progression of CWB fatigue cracking observed by the manager of the C-130 Aircraft Structural Integrity Program (ASIP) and Lockheed relative to the C-130 EBH cohort groups. Below 20,000 EBH, CWBs are generally crack free or have very small cracks below the threshold of detection. Beyond the 20,000-EBH threshold, cracks begin to be detected and can usually be repaired. Below 30,000 EBH, single cracks usually appear at well-known fatigue-critical locations. Beyond 30,000 EBH, cracking becomes more widespread and is not limited just to fatigue-critical locations, and multisite damage (MSD) and multielement damage (MED) begin to appear. Given the amount of accumulated center wing EBH, most C-130Es and H1s would be expected to exhibit MSD and MED already.

Figure 2.8
Progression of Center Wing Box Cracking with Usage

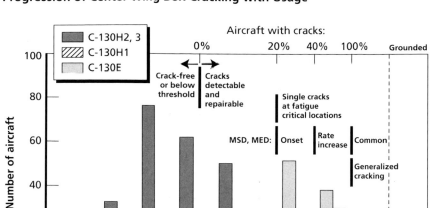

SOURCES: CWB sheet, January 2007; LMAC, 2006;
WR-ALC, 2005.
NOTE: Depicts unretired aircraft.
RAND *MG818-2.8*

The prevalence of cracks at fatigue-critical locations and the presence of MSD and MED markedly increase as EBH rises from 30,000 to 35,000 and do so even more as aircraft approach the 38,000-EBH flight restriction threshold and the 45,000-EBH grounding limit.[12] With typical air mobility aircraft accumulating 1,000 or so CWB EBH per year, only limited time remains to deal with the fatigue that aging E and H1 models are experiencing before they pass the aforementioned thresholds.

The number and severity of fatigue cracks found at fatigue-critical locations of the CWB by depot- and field-level inspections from 2001 to 2004 exceeded predictions. Inspections found 123 aircraft with cracks at fatigue-critical locations of the center wing. The prevalence

---

[12] WR-ALC, 2005.

of MSD and MED raised further concerns. More recently, long cracks have appeared at much lower EBHs than fatigue-cracking models predicted (see the 1.8-inch crack that appeared at 29,000 EBH in Figure 2.9, which is shown by the red dot off the top of the chart).[13] These "outliers," together with the inherent scatter in the distribution of service cracks apparent in the figure, raise questions about the precision of programs for tracking the usage and structural health of C-130 airframes. The accelerated cracking that occurs at higher EBHs (see the green durability limit curve in Figure 2.9) and the sizable population of aging C-130s in high-EBH cohort groups (refer to Figures 2.7 and 2.8) have heightened Air Force management concerns about flight safety and overall CWB structural health.

In September 2004, the Air Force formed the Center Wing Independent Review Team (IRT) to assess the risks fatigue cracking at the three center wing fatigue zones poses (see the appendix) and to recommend actions. One month later, the Air Force imposed flight restrictions on 43 high-time aircraft. The team addressed each of the zones in meetings through 2006. It called for the imposition of flight restrictions at 38,000 EBH,[14] grounding at 45,000 EBH, a thorough depot-level inspection and repair of the lower wing surface to meet Time Compliance Technical Order 1908 (TCTO 1908) to lift flight restrictions to 45,000 EBH,[15] and development of a center wing replacement

---

[13] Marian Fraley and Peter Christiansen, "C-130 Groundings and Restrictions, WR-ALC, 330 ACSG," briefing to RAND, U.S. Air Force, Program Analysis and Evaluation, Rosslyn Va., October 20, 2006.

[14] Restrictions were expressed in terms of a number of physical and flight parameters to reduce maximum wing up-bending loads to below 60 percent of the design limit at 38,000 EBH (G. R. Bateman and P. Christiansen, "C-130 Center Wing Fatigue Cracking, A Risk Management Approach," presented at the 2005 U.S. Air Force Aircraft Structural Integrity Program Conference, Memphis, Tenn., November 29–December 1, 2005).

[15] A mixture of nondestructive inspection (NDI) techniques provides nearly 100 percent NDI of the CWB lower surface at fastener holes and notches. Spar webs are also inspected. Some critical areas must undergo double independent inspections to reduce the risk of missing cracks. Inspectors use bolt-hole eddy-current probes, eddy-current surface scans, and a magneto-optic imager. G. R. Bateman, "Wing Service Life Assessment Methodology & Results," presented at the 2005 Hercules Operators Conference, Atlanta, Ga., October 2005; Peter Christiansen, WR-ALC, "Assessment of U.S. Air Force Center Wing Cracking,"

Figure 2.9
Inspections Have Found Some Surprisingly Long Cracks at Comparatively Low Equivalent Baseline Hours

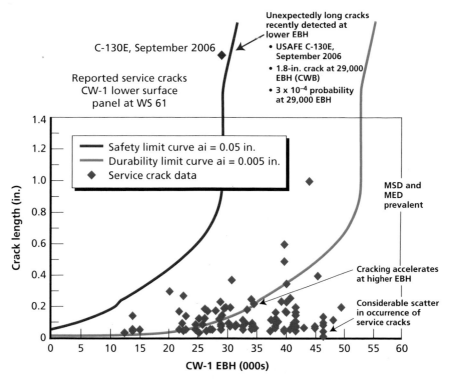

SOURCE: Fraley and Christiansen, 2006.
NOTE: Unexpected cracks and data scatter may in part reflect imprecision in tracking the structural health of the airframes of individual aircraft.
RAND *MG818-2.9*

program. The IRT also recommended implementation of an updated flight-load program to characterize the severity of current aircraft usage, identified problems associated with the reliability of aircraft inspections, and recommended actions to mitigate the risks that cracking in the rainbow fittings poses.[16]

---

presented at the 2006 Hercules Operators Conference, Atlanta, Ga., October 2006; Fraley and Christiansen, 2006.

[16] Fraley and Christiansen, 2006.

Recognizing the prevalence of MSD and MED in the CWB, Lockheed moved from a risk-assessment methodology assuming a single dominant fatigue crack scenario to an advanced analytic technique that assumes an MSD crack scenario. A Monte Carlo simulation determines the time it takes for MSD cracks to link up to develop a probability distribution of crack length as a function of EBH. This constitutes a key input in the derivation of the single flight probability of failure as a function of EBH.[17] Figure 2.10 shows a fall 2006 result from this methodology.[18]

The Air Force has used Lockheed's MSD risk-analysis methodology together with definitions of acceptable flight risks from the ASIP, Military Standard 1530C (MIL-STD-1530C), shown as color bands in Figure 2.10, to set policies for imposing operating restrictions and grounding actions. Cracks link up as multisite cracking increases with EBH, causing the residual strength of the CWB structure to decrease. This increases the probability that the stresses caused by a gust or maneuver will exceed the design limit load of the structure, resulting in a structural failure. When the single flight probability of failure reaches the MIL-STD-1530C threshold of acceptable risk, 1×10^{-7}, estimated to occur at approximately 38,000 EBH using the Lockheed methodology for typical AMC usage (green curve in Figure 2.10), the Air Force imposes flight restrictions to reduce risks.[19] Flight restrictions reduce loads and the rate of crack growth, but ultimately, as cracks continue to grow, the Air Force grounds aircraft at 45,000 EBH, when the single flight probability of failure is estimated to reach the 1×10^{-7} threshold again (blue curve in Figure 2.10).

Figure 2.11 shows results from a more-recent refinement to Lockheed's risk analysis methodology, reported at the December 2006 ASIP

---

[17] Christiansen, 2006.

[18] Figure 2.9 depicts risk curves for more-damaging special operations flights, typical air mobility flights, and flights with operating restrictions imposed to reduce the single-flight probability of failure. A subsequent figure will depict the effect of the TCTO 1908 inspection and repair on the risk curve.

[19] Christiansen, 2006; MIL-STD-1530C, *Aircraft Structural Integrity Program (ASIP)*, Wright-Patterson AFB, Ohio: Aeronautical Systems Center, November 1, 2005.

Figure 2.10
Risks Increase with Equivalent Baseline Hours as Multisite and Multielement Damage Reduce Residual Strength

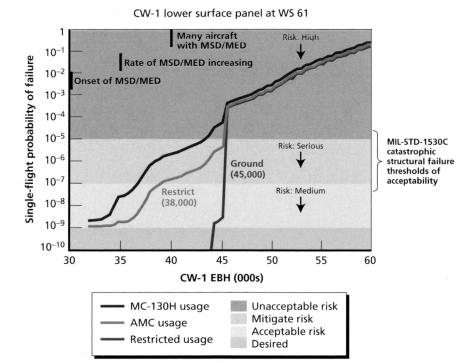

SOURCES: Fraley and Christiansen, 2006; G. R. Bateman, "Wing Service Life Analysis Update," presented at the 2006 Hercules Operators Conference, Atlanta, Ga., October 2006; MIL-STD-1530C, 2005.
RAND MG818-2.10

conference, that illustrate how a major inspection can temporarily reduce flight risks to the 1×10^{-9} threshold, allowing unrestricted operations to higher EBHs.[20] The major inspection (TCTO 1908) and associated repairs reduce risks, but limitations in the reliability of inspections and MSD ultimately cause risks to rise again.

---

[20] Successive refinements to the MSD risk analysis show some differences in estimates of when risk thresholds are breached, although the general character of the risk curves are similar. The Air Force has thus far retained the original 38,000-EBH and 45,000-EBH thresholds for flight restrictions and groundings. The figure depicts risks for AMC combat delivery usage (Christiansen, 2006).

Figure 2.11
Failure Risks Increase Dramatically at Higher Equivalent Baseline Hours

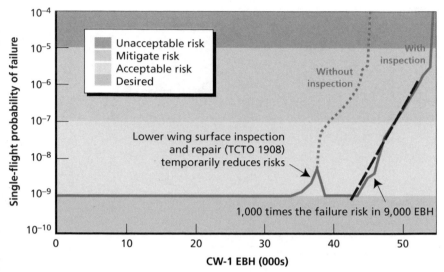

SOURCES: P. Christiansen, G. R. Bateman, and A. Navarrete, "C-130 Center Wing MSD/MED Risk Analysis," presented at the 2006 U.S. Air Force Aircraft Structural Integrity Program Conference, Memphis, Tenn., November 28–30, 2006; MIL-STD-1530C, 2005.
NOTE: LMAC MSD model risk assessment.
RAND *MG818-2.11*

Note the steepness of the risk curve at high EBH in the logarithmic plot of Figure 2.11. The Lockheed methodology shows that risk increases by three orders of magnitude (from 1×10^{-9} to 1×10^{-6}) in just 9,000 EBH (see dashed red line).[21] In the appendix, we will illustrate some of the sources of uncertainty in accurately estimating C-130 usage and its associated structural health over time. Figure 2.11 shows that underestimating the severity of past usage could mean flying aircraft at higher-than-acceptable risk. Being wrong about the estimated EBH of an aircraft could have significant consequences, as could delib-

---

[21] RAND has not performed an independent assessment of this MSD risk-analysis methodology. A sensitivity analysis using a somewhat simpler risk-analysis formulation for the same fatigue-critical location shows the general character of the results to be quite robust to changes in key parameters.

erately deciding to fly with higher risk beyond the recommended EBH grounding threshold.[22]

Aircraft grounded before reaching the CWB-grounding threshold of 45,000 EBH are one manifestation of uncertainty in the structural health of the C-130 fleet. After inspecting and finding considerable CWB cracking, operators have decided to not repair 12 C-130E aircraft having less than 45,000 EBH (see Figure 2.12), including an aircraft listed as having only 33,700 EBH.[23] The population of so-called

Figure 2.12
Twelve Aircraft Have Been Grounded Before Reaching 45,000 Equivalent Baseline Hours Because of Extensive Center Wing Box Fatigue Damage

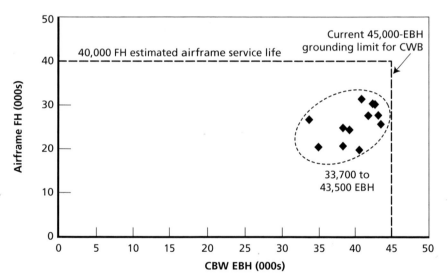

SOURCES: C-130 SPO, CWB data from July 2006, January 2007, and March 2007.
RAND *MG818-2.12*

---

[22] Chapter Six will illustrate the potential consequences, measured in terms of aircraft losses, from flying beyond the recommended EBH grounding threshold.

[23] Decisions to ground the 12 aircraft prior to 45,000 EBH have mostly been framed in economic terms. Except for an anecdotal report that one aircraft may be retired in place rather than incur the risk of even one ferry flight to another location, RAND lacks details about the flight risks the 12 aircraft faced prior to grounding due to fatigue damage.

red-X grounded aircraft has grown during the course of this FSA, and the Air Force has already officially retired five of these aircraft.[24] An Air Force decision to begin grounding aircraft prior to 45,000 EBH as a matter of policy to control risks, rather than on a case-by-case basis, would have significant implications for the C-130E/H inventory. Figure 2.13 illustrates the consequences of different grounding thresholds for the inventory of air mobility C-130Es and C-130Hs. Since Air Force policy currently calls for grounding aircraft at 45,000 EBH, only thresholds below 45,000 are germane. The reduction in the number of aircraft is significant. For example, grounding aircraft at 35,000 EBH rather than 45,000 EBH would eliminate about 80 aircraft from the operational inventory by 2017 (Figure 2.13). Further examination of Figure 2.13 shows that these groundings could reduce the C-130E and H inventory by 18 to 25 percent over time. Filling the resulting inventory gap would require an additional multibillion-dollar procurement of replacement aircraft, should that be the policy option chosen to make up the difference.

Although the current experience with grounded aircraft prompted us to explore the implications of lower grounding thresholds, other factors might also change grounding policies. Some of these could include updated operational load measurements that could show that the flight environment has become more severe or discovery of unexpected corrosion problems and the results of the outer-wing risk assessment completed in early 2009.[25]

Other System Issues

Other C-130 systems besides structures will require attention as the aircraft age. A clear vision of the time phasing of all the investments required to keep aircraft viable can inform decisions about whether

---

[24] CWB spreadsheets for July 2006, January 2007, and March 2007.

[25] LMAC completed the risk assessment under contract to Warner Robins in early 2009. The results of the analysis were consistent with the full-scale wing-durability results; risk remains acceptable to the 60,000 EBH service life.

Figure 2.13
Uncertain Timing of Aircraft Grounding or Repairs Could Greatly Affect Inventory

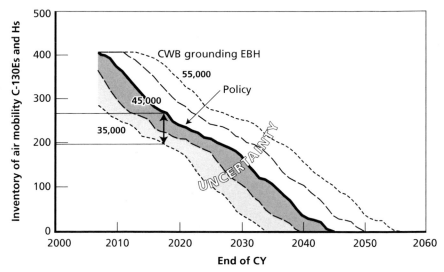

SOURCES: CWB sheet, July 2006; CWB sheet, January 2007; CWB sheet, March 2007; Fraley and Christiansen, 2006.
NOTE: Scatter reflects natural fatigue phenomena but is also a symptom of possible problems measuring usage. Twenty-three aircraft have "passed" TCTO 1908, with restrictions removed to 45,000 EBH, but 12 red-X groundings have occurred from 33,700 to 43,500 EBH.
RAND *MG818-2.13*

to retain or retire aircraft (see Figure 2.14).[26] For the FSA, Lockheed showed RAND briefing materials prepared for the Air Force Fleet Viability Board's review of C-130E and H1 aircraft.

In addition to structural issues, Lockheed identified a near-term need for cockpit upgrades to navigation, communication, and safety and surveillance systems so that C-130s can comply with Global Air Traffic Management standards and not be restricted to slower air routes and low altitudes. Older aircraft also have significant wiring problems caused by fatigue, vibration, repeated manipulation of wiring during

---

[26] It is not clear that the Air Force has invested sufficiently in sustaining engineering to have a clear vision of time-phased modification requirements and their costs for all the systems on C-130s.

**Figure 2.14
Aging Functional Systems Will Compete with Structures for
Modification Resources**

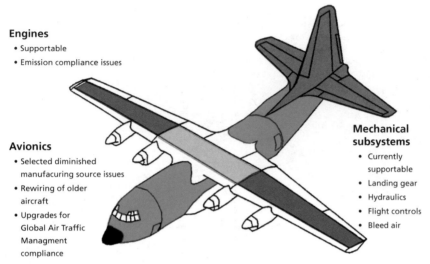

Engines
- Supportable
- Emission compliance issues

Avionics
- Selected diminished
  manufacuring source issues
- Rewiring of older
  aircraft
- Upgrades for
  Global Air Traffic
  Managment
  compliance

**Mechanical
subsystems**
- Currently
  supportable
- Landing gear
- Hydraulics
- Flight controls
- Bleed air

SOURCE: Adapted from LMAC, 2006.
NOTE: Time-phased modification requirements for structures and functional
systems are needed to make informed decisions about retention and retirement.
RAND MG818-2.14

maintenance, and exposure to heat, chemicals, etc. Some operators
are rewiring their C-130s, but the Air Force would need to evaluate
the appropriateness of rewiring its high-EBH aircraft in the context
of whatever other life-extension modifications it plans to accomplish.
Lockheed further identified diminishing manufacturing source issues
for selected avionics systems.[27]

For the most part, Lockheed assessed most other systems as currently supportable. However, any study comparing materiel solutions,
including new aircraft buys and SLEPs, will need to assess in more
detail the succession of aircraft systems that will require attention as
C-130s age.

The SLEP for Royal New Zealand Air Force C-130Es provides
an example of the scope of work involved in a major life-extension pro-

---

[27] LMAC, 2006.

gram. That work, being performed by Spar Aerospace Limited (part of L3 Communications), includes extending structural life through replacement and refurbishment, fatigue monitoring and fatigue improvements, reliability improvements (including complete rewiring), mechanical system upgrades, next-generation cockpit upgrades, and preventative maintenance programs.[28]

Should the Air Force decide to undertake a SLEP for its C-130s, specific efforts to extend structural life will constitute an important part of the SLEP. Chapter Six will outline some of the options available for extending the structural life of C-130s.

---

[28] Asad Baig, Spar Aerospace Limited, "Royal New Zealand Air Force C-130 Life Extension Program," presented at the 2005 Hercules Operators Conference, Atlanta, Ga., October 2005.

CHAPTER THREE

Functional Solution Analysis Methodology

This chapter describes the methodology used in the FSA to evaluate potential doctrine, organization, training, materiel, leadership and education, personnel, and facilities (DOTMLPF) and policy options to close the emerging shortfall in intratheater airlift capability identified in the prior FNA.[1]

Current and Projected Fleet Sizes

As discussed in Chapter Two, MAF included 405 C-130E/Hs (TAI) on January 3, 2007. The Center Wing EBH Report implicitly projects an annual flying-hour rate, as well as an annual EBH accumulation rate, for each of these 405 C-130E/Hs.[2] The report explicitly includes, for each of the 405 C-130E/Hs, the level of cumulative EBH as of January 3, 2007; the projected date when the aircraft will have accumulated 45,000 EBH; and the number of flying hours it will have accumulated by that date. Annual flying-hour and EBH rates follow from those.

---

[1] Stillion, Orletsky, and Rosello, 2011.

[2] The Center Wing EBH Report is a spreadsheet drawn from the C-130 System Program Office's AIRCAT database on January 3, 2007, that provides EBH to date and projects when each C-130's CWB will reach 38,000 EBH and 45,000 EBH, based on historical flying patterns. This report was the basis for our fleet-life projections, and we therefore refer to it specifically by name throughout this monograph to eliminate confusion with other reports drawn from the same database (primarily in Chapter Two).

The Air Force also had 37 MAF C-130Js on January 3, 2007. Therefore, the size of the current MAF fleet is 442.[3] In addition, recent budget documents project that the Air Force will acquire an additional 28 MAF C-130Js by the end of FY 2010.[4] Of course, actual procurements are likely to turn out somewhat different as future sessions of Congress rework budgets. For this analysis, we took the projection of 28 additional C-130Js as a given. The results of this study will hold unless this number changes significantly from the budget projection. If the number of aircraft procured turns out to be significantly different, the inventory drawdown curves presented in this document will offer a sense of the change in the year the inventory will fall below the MCS requirement (see Figure 3.1).

Figure 3.1 shows the drawdown curve for the C-130 fleet using each aircraft's unique annual flying-hour and EBH rates, as projected

Figure 3.1
Number of C-130s in MAF Inventory and MCS Requirement

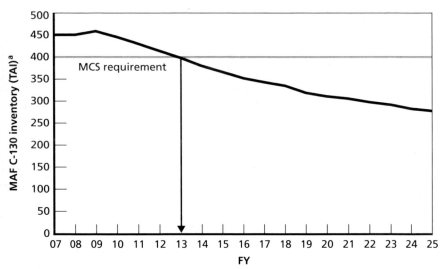

[a]Inventory numbers assume that all aircraft undergo TCTO 1908 inspection and are able to fly 45,000 EBH.

RAND *MG818-3.1*

---

[3]   Some of these 442 C-130s are in restricted flying status.

[4]   The projection is based on U.S. Air Force, 2007b.

in the Center Wing EBH Report. This figure shows that the number of C-130s will fall below the MCS requirement of 395 in 2013.[5] In the analysis, each aircraft can contribute to the requirement until it reaches 45,000 EBH, at which point it is retired. We further assumed that all aircraft would undergo and successfully complete TCTO 1908 at 38,000 EBH, allowing them to fly to 45,000 EBH unrestricted. Even with these optimistic assumptions, this figure shows that the number of C-130s will be below MCS requirement in 2013. If some aircraft cannot be repaired after undergoing the inspection and repairs outlined in TCTO 1908 or if some aircraft must undergo TCTO 1908 prior to the 38,000-EBH mark, the shortfall could occur prior to 2013.[6]

It is interesting to note that there is a significant difference in the remaining life for aircraft in the active and reserve components. Figure 3.2 shows the years remaining for both components. The oldest aircraft are primarily in the active component, while the majority of the newer aircraft are in the reserve component.

Analysis Methodology

The first step in the FSA was to identify potential materiel and nonmateriel solutions that might be able to close the capability gap identified in the FNA. We identified a number of potential solution options for each of the DOTMLPF categories, then described each and the effects of a variety of factors relating to its suitability. The factors included technical and operational risk, supportability, and effects on other systems. In addition, we identified previous analyses and/or data sources

---

[5]   A new requirement of 335 C-130s is defined in *Mobility Capabilities & Requirements Study 2016*, which was released after the completion of this work. Figure 3.1 indicates that the fleet will fall below the 335 requirement in 2017. (DoD, *Mobility Capabilities & Requirements Study 2016*, Washington, D.C., February 26, 2010, Not Available to the General Public.)

[6]   Successful completion and repair of an aircraft under TCTO 1908 does not guarantee the ability of the aircraft to reach 45,000 EBH. TCTO 1908 allows an additional 7,000 EBH after completion of the inspection and repairs. In addition, since the repair process involves reworking rivet holes, TCTO 1908 can only be conducted once on an aircraft. An aircraft that undergoes TCTO prior to 38,000 EBH will need to be grounded before it reaches 45,000 EBH. (AMC/A4M provided this important clarification of TCTO 1908.)

Figure 3.2
Projected Service Life Remaining for Active- and Reserve-Component C-130s

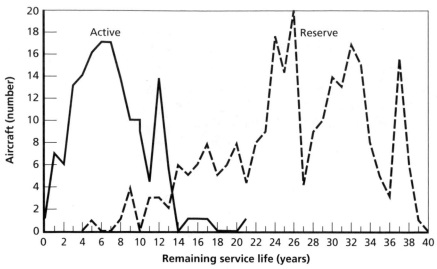

that could be used to analyze each option. An integrated product team (IPT) then convened, to which this set of options and initial preliminary analysis was presented. The IPT included representatives from the Air Staff, AMC, OSD/PA&E, the Army, and the contractor that had conducted the Army's Future Cargo Aircraft AoA. Members of the IPT provided feedback on the options RAND had presented. The IPT then had the opportunity to suggest and discuss other potential solution options. This meeting helped us develop a set of options for analysis during the FSA.

We then assessed the potential for each of these options to close the capability gap, identified potential negative implications of implementing each option, and presented this preliminary analysis to a second IPT for review. This analysis and the feedback from the IPT meeting helped us identify a subset for a more-detailed analysis that we presented to the sponsor in an interim project briefing. The more-detailed analysis involved quantifying the potential effects of each solution option on C-130 fleet life and the potential for each to close the capability gap. This led to a set of options for a net-present-value (NPV) cost analysis.[7] The screening process and the detailed analysis used a variety of input information, including the MCS C-130 requirements, the current fleet mix, the expected drawdown due to fatigue and corrosion, and cost estimates. Figure 3.3 graphically represents the analytical process we used to conduct this FSA.

Figure 3.3
The Functional Solution Analysis Process

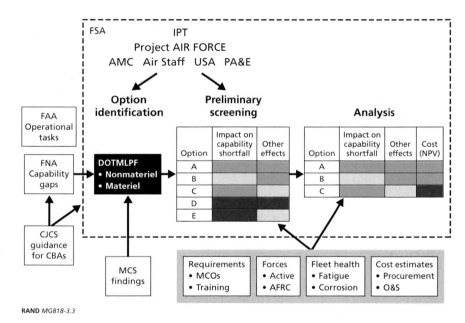

RAND MG818-3.3

---

[7] Not all options that received a detailed effectiveness analysis were costed. We did not cost all the options because the detailed effectiveness analysis revealed some concerns that were not apparent in the initial screening that rendered them nonviable. For example, although we had concerns during the initial screening process, we further explored the possibility of reducing crew qualifications. We were not able to identify a way to reduce crew qualifications without significantly affecting capability, and the potential to close the capability gap was minimal. We therefore did not cost this option.

Three Classes of Options

After we identified the options during the first IPT, three broad classes emerged. We use these as collectors to organize options:

- reducing the rate of accumulation of fatigue damage (measured in terms of EBH) of the current C-130 fleet
- increasing the supply of EBH
- meeting the requirement with fewer C-130s.

Delaying the Need to Recapitalize—Where's the Leverage?

The potential leverage that can be achieved by implementing an option from given class varies considerably. We made an arbitrary parametric change of 25 percent in each class and measured the effect on the required C-130 recapitalization date. Although such changes would not necessarily be possible, our parametric analysis does suggest the amount of leverage—in terms of delaying the need to recapitalize the fleet—implementing changes in each of these classes could achieve. The understanding this analysis provides informs the discussions of individual options later in this monograph.

Figure 3.4 shows that a 25-percent reduction in the EBH accumulation rate could delay the need to recapitalize by only about two years, to 2015, because so many C-130s are already close to retirement. Potential ways to reduce EBH accumulation include additional use of simulators and additional use of companion trainer aircraft (CTA). For example, an aircraft that has three years of life remaining under current EBH usage patterns could gain only one additional year with a 25-percent reduction in EBH usage. The high EBH accumulations of so many C-130s, already putting them close to the EBH limit, mean that fairly significant reductions in EBH usage have limited potential to delay the need to recapitalize.

Figure 3.5 shows that an arbitrary 25-percent increase in the amount of EBH available for each aircraft prior to grounding could delay the need to recapitalize by about nine years, to 2022. However,

Figure 3.4
Reducing Accumulation of Equivalent Baseline Hours by 25 Percent

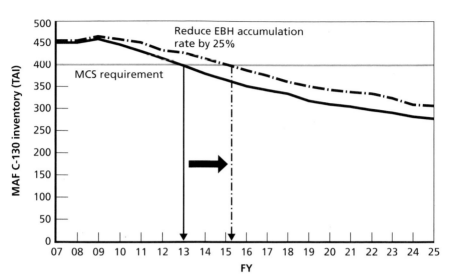

NOTE: Ways to reduce usage rates include the use of simulators, companion trainers, and C-17 substitution.

RAND *MG818-3.4*

flying an aircraft after 45,000 EBH could entail a significant risk of flight failure unless steps have been taken to mitigate structural fatigue damage beyond that limit.[8]

Figure 3.6 shows that meeting the requirement with fewer C-130s can also provide significant leverage. If the MCS requirement could be met with 25 percent fewer C-130s, the recapitalization need could be delayed by about eight years, to 2021.

Figure 3.7 is presented for completeness, to reflect how SLEPs or new buys affect the fleet. In this case, the fleet is allowed to draw down to the requirement. Then, the gap is filled with SLEPed or new-buy aircraft to maintain a fleet of 395 TAI aircraft. For our purposes,

---

[8] The cases shown here give a sense of the leverage broad categories of policy solutions offer for addressing the capability gap. Increasing the EBH limit on each aircraft by 25 percent to over 56,000 EBH involves a great deal of risk. Later in this monograph, we will show this risk is unacceptably high unless significant modifications are undertaken to mitigate CWB structural fatigue issues.

Figure 3.5
Increasing Availability of Equivalent Baseline Hours by 25 Percent

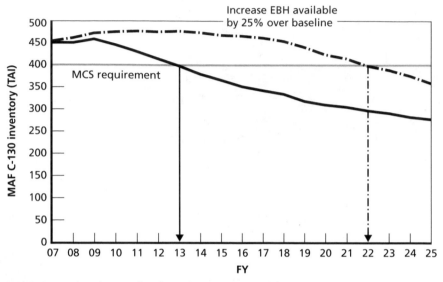

NOTE: Increasing the supply of EBH increases EBH tolerance.

Figure 3.6
Decreasing Number of C-130s by 25-Percent

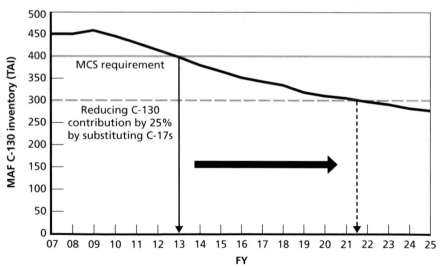

Figure 3.7
Increasing Availability of Equivalent Baseline Hours

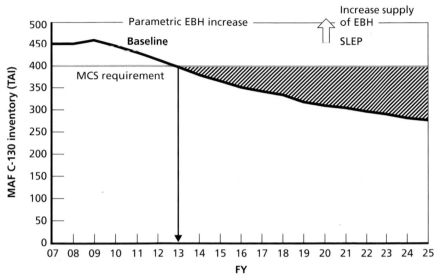

RAND *MG818-3.7*

major SLEPs and new buys are materiel solutions, and this monograph therefore does not address them fully.[9]

In the next three chapters, we evaluate specific options to close the capability gap. Each chapter covers one of the three broad classes of options.

---

[9] As discussed earlier, FSAs focus on nonmateriel solutions. Since our analysis found no viable nonmateriel solution, SLEPs are included in the subsequent UIAFMA (Kennedy et al., 2010), along with new aircraft procurement programs. The UIAFMA conducted a cost-effectiveness analysis to determine the best approach to recapitalize the intratheater airlift fleet. This analysis considered a variety of new aircraft alternatives, as well as service-life extensions of the older C-130 aircraft.

CHAPTER FOUR

Delaying C-130 Recapitalization by Reducing Accumulation of Equivalent Baseline Hours

This chapter explores the potential for delaying the need to recapitalize the C-130 mobility airlift fleet by reducing the rate of fatigue damage accumulation, as measured by EBH. The methodology and analysis focus on more than simply reducing flight hours. Rather, we target the flying hours that most stress the C-130 airframe, specifically the CWB. Table 4.1 shows the 11 options considered for reducing EBH accumulation.

Before exploring the potential solution options, the effects of different types of flying on EBH accumulation need to be understood. The relative severity of different flying events highlights where there is leverage to reduce the fatigue on the airframe. As discussed earlier, EBH is the product of flying hours and a severity factor that depends on the characteristics of the sortie. For example, if low-level training missions have an average severity factor of 6 and if channel missions have a severity factor of 1, eliminating one hour of low-level flying saves six times the EBH that eliminating one hour of channel flying would. Channel missions deliver cargo and personnel throughout the world. These are typically characterized by long sortie durations whose origins and destinations are well-developed airbases. These missions put the least stress on the airframe.

The next section addresses the relative severity of the different flying missions. In addition, a brief discussion of the peacetime flying-hour program provides an understanding of the types and amounts of flying required to sustain the readiness of the C-130 crew force. A

Table 4.1
Options for Reducing Equivalent Baseline Hour Accumulation

Options	Estimated Potential Impact	Other Implications
Most promising		
Increase use of simulators for training	High	None
Increase use of CTA	High	None
Shift high-severity-factor operational missions to other aircraft	High	None
Reduce crew qualifications	Moderate	Loss of capacity or flexibility
Reduce high-severity-factor training	Moderate	Loss of capacity or flexibility
Dropped in the screening process		
Rotate aircraft among components	Moderate	May not be viable
Increase experience mix	High	Effects on personnel
Change active-reserve mix	Moderate	Few active units Effect on temporary duty (TDY)
Add ANG/AFRC associate units to active squadrons	Low	Crew ratio cuts needed
Increase squadron size	Very low	Reduced flexibility
Place flight restrictions on specific aircraft	Very low	Reduced flexibility

Key:
Green, *few* or *none*
Yellow, *moderate*
Red, *significant*

detailed analysis of the promising solution options follows the flying-hour program explanation. The potential of each solution option to delay the need for recapitalization was estimated by first determining the flying-hour savings. From the flying-hour savings, a percentage decrease in EBH accumulation was derived to calculate the increased fleet life. The final section of this chapter presents the rationale for screening out the proposed solution options that did not receive detailed analysis. For each of these solution options, information is presented that briefly highlights either the lack of EBH savings, significant barriers to implementation, or other negative consequences of these options.

How Different Types of Flying Relate to Accumulation of Equivalent Baseline Hours

EBH accumulation can be reduced through a combination of two general approaches, flying the aircraft less and flying the airplane in a manner that puts less stress on the airframe. Understanding the amount of savings that each approach yields requires understanding the relative severity of different C-130 missions.

C-130s fly many different missions around the world. These missions vary widely in the amount of stress they put on the airframe. As discussed in Chapter Two, a severity factor is defined for each mission that is then multiplied by the number of flight hours to determine the EBH for that mission. An example of a low-severity-factor mission would be a sortie to ferry the aircraft between two locations. For this mission, the EBH may actually be less than the flight time. In contrast, a training mission combining high-speed, low-level training, assault landings, and instrument-approach training will stress the airframe more heavily—potentially resulting in a severity factor of three or four.

Flying is tracked differently by the operational and the engineering communities. The mobility operational community generally tracks flight hours by funding source. Beyond funding source, flight hours are broken down into categories that describe the payment arrangement or type of flying in more detail. In contrast, the engineering community, responsible for the fleet's structural health, tracks flying by the specific events accomplished on each sortie. Either method alone is not sufficient for determining the effect of reducing flying hours or types of flying on the C-130 fleet's structural life. The operational method, while useful for examining funding streams and budget planning, provides no insight into the damage accumulated. Additionally, the same operational mission classification may have individual sorties that span several of the engineering categories. On the other hand, the engineering categories provide detailed damage accumulation but do not indicate a sortie's purpose, making it difficult if not impossible to analyze the EBH savings from the different proposed solution options.

The air mobility community uses the Global Decision Support System (GDSS) to track the different types of flying mission performed by mobility aircraft worldwide. The mission types in GDSS are contingency, channel, training, joint airborne and air transportability training (JA/ATT), special assignment airlift missions (SAAMs), and guardlift.

Each GDSS mission type serves a different purpose and is funded differently. Contingency missions are typically in direct support of an ongoing operation and are usually funded specifically from a war or contingency allocation. Channel missions are regularly scheduled routes flown to DoD locations worldwide and are funded by "user" agencies that pay set rates depending on the origination and destination locations, the weight of cargo, and the number of passengers. Training missions are flights to increase the readiness and proficiency of the aircrews. JA/ATT missions are training missions that provide the Army a platform for paratroop training and cargo loading. In SAAMs, DoD and other users pay for the use of the entire airplane by flight hour for a specific mission. Guardlift missions are airlift missions serving and funded by the National Guard in support of Army and ANG unit deployments. These six mission types made up over 98 percent of all the C-130 sorties flown in 2005–2006.

The C-130 System Program Office uses the AIRCAT information system to track and catalog flight time, EBH accumulation, and various other flying and structural component histories for the C-130 fleet. AIRCAT groups missions into three main classifications: training, logistics, and low-level sorties. AIRCAT differs from GDSS in classifying only sorties during which the aircraft performed touch-and-go landings as training missions, and missions in which no touch-and-go landings or low-level flying took place as logistics missions. Low-level sorties are those having high-speed, low-level flight times exceeding 15 minutes. The program office assigns a specific EBH increment to each sortie according to which of 1,621 mutually exclusive profiles the sortie matches. Factors used to determine the sortie profile include duration; number of touch-and-go landings; number of stop-and-go landings;

low-level, high-speed flight time; average altitude; takeoff fuel weight; and cargo weight.

Since GDSS categorizes sorties by mission type and AIRCAT provides the EBH accumulated during each sortie, we combined sortie-level data from both databases to quantify EBH accumulation by mission type. Matching these allowed us to determine average severity factors for different types of GDSS mission categories. Using data from CYs 2005 and 2006, we matched over 125,000 sorties representing approximately 75 percent of the total recorded sorties in the AIRCAT database.[1] Table 4.2 presents the average severity factors for the GDSS mission categories. Figure 4.1 presents the percentage of total sorties, flying hours, and EBH for each of these mission categories.

The C-130 Flying-Hour Program: Understanding Where Flight (and Equivalent Baseline) Hours Can Be Reduced

To understand what flying hours can be saved from the C-130 fleet, it is important to understand how the flying-hour program is planned and developed. The Air Force bases its annual flying-hour program for the different aircraft on peacetime home-station flying requirements to

Table 4.2
Average Severity Factor by
GDSS Mission Type

GDSS Mission Type	Average Severity Factor
Training	2.61
JA/ATT	2.52
Contingency	1.80
Channel	1.08
SAAM	1.06
Guardlift	0.99

[1] We assumed that the sorties not matched had the same distribution as the matched sorties.

Figure 4.1
Relative Sorties, Flying Hours, and EBH of C-130 Fleet

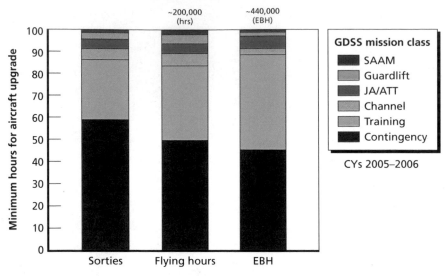

RAND *MG818-4.1*

maintain the readiness of the aircrews, maintainers, and other DoD elements.

The general Air Force methodology for developing a flying-hour program starts with the number of primary aircraft authorized (PAA) and factors in aircrew data and training requirements to arrive at the total number of flying hours. Aircrew data that go into the flying-hour calculations may include crew ratio, ratio of inexperienced to experienced pilots, number of pilots in staff positions, pilot-production numbers, and the number of in-unit upgrades and qualification courses. Training requirements include such items as mission qualification training, proficiency training, pilot seasoning, number and frequency of specific training events, operational mission, and event refly rates.

In the MAF, two factors dominate the flying-hour program: pilot seasoning and event-based training. *Seasoning* refers to the process of turning new graduates from undergraduate pilot training (UPT) into aircraft commanders. Since this requires 700 hours of flying in the C-130, and a normal tour lasts 28 months, this translates into a seasoning rate of 25 hours per month. Because 90 percent of the pilots entering the C-130 crew force come directly from UPT, the seasoning rate is a significant driver of the flying-hour program.

Event-based pilot training consists of the specific events that pilots must perform on a regular basis over the course of a year. For example, all pilots must perform at least one takeoff, instrument approach, and landing each month. There are many events, each to be accomplished a specific number of times per semiannual period. Other events include low-level training, night-vision goggle landings, tactical arrivals, tactical departures, and engine-out takeoffs and landings.[2] Required training is detailed by pilot experience, crew position (i.e., copilot, aircraft commander), and mission qualification (i.e., formation airdrop).

The active-duty AMC flying-hour program will be used as an example of how flying hours are programmed for the C-130 fleet. An "experience ratio" is applied across the fleet to account for delays in attending aircraft commander upgrade training. The ratio of inexperienced pilots to experienced pilots is 43:57. To get the number of hours planned, use the following formula:

$$Hours = PAA \times Crew\ ratio \times \left(\frac{Pilots}{Crew} \right)$$
$$\times Percentage\ of\ inexperienced\ pilots \times Seasoning\ rate.$$

After determining the total number of hours needed for seasoning, the different types of required flying are programmed against the total. First, all the event-based pilot training hours are totaled. This summation is accomplished by assigning each of the required events a fixed duration. For example, an instrument approach is assigned a duration of 0.3 hours. The total time for all the pilots for event-based training is then calculated as follows:

$$Total\ time = Number\ of\ pilots \times Number\ of\ events$$
$$\times Event\ duration.$$

---

[2] The events and required frequency of accomplishment are detailed in the training tables found in Air Force Instruction (AFI) 11-2C-130, *Flying Operations*, Vol. 1: *C-130 Aircrew Training*, Washington, D.C.: Department of the Air Force, July 19, 2006.

In addition to the event-based training, the C-130 crew force must accomplish a significant amount of mission training. This mission-based training consists primarily of airdrop training with the Army, which is used to train and keep qualified the Army's Airborne forces. Additionally, Red Flag exercises make up some of the required mission-based training. Note that the event-based training and the mission-based training are not additive; credit is given for the training events accomplished during the mission-based training. For example, on every JA/ATT sortie during which Air Force crews drop Army parachutists, the crew also accomplishes such events as a takeoff, a landing, low-level flying, and an airdrop. These events are credited toward the event-based training requirements, thus reducing the required hours determined from a straight summation of the training events and durations.

For the active component, event-based and mission-based training activities provide approximately 60 percent of the required flight hours needed to season inexperienced pilots. The remaining 40 percent are "user hours," gleaned when the C-130 provides actual airlift for other DoD and government users.

The Air Force funds mission- and event-based training hours using operations and maintenance dollars. The various agencies that "hire" the C-130 for airlift service pay for the user-funded hours through the Transportation Capital Working Fund (TWCF).

Flying-hour programming methodology is very similar for the other active-duty major commands, the ANG, and the AFRC. The ANG and AFRC have fewer inexperienced pilots; hence, the training burden for the Air Reserve Component (ARC) is less than that for the active component. ARC members typically have already been trained and become experienced during previous time on active duty. However, the ARC applies the same methodology in building its flying-hour programs.

With an understanding of the general construction of a flying-hour program, solutions can be targeted to specific areas of C-130 flying and then related to the amount of EBH accumulation prevented. Figure 4.2 presents the AMC active-duty flying-hour program for FY 2006.

Figure 4.2
Example of Flying-Hour Program Build: AMC Active-Duty Annual
Programmed Flight Hours for FY 2006

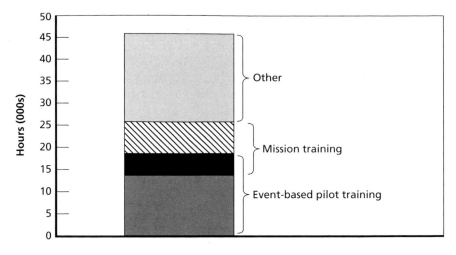

Options Analyzed

The first part of this section focuses on options that received a detailed effectiveness analysis. Then, it discusses options we eliminated during the initial screening process, along with the rationale for their elimination. Five potential solution options survived the initial screening process and were analyzed in greater detail: greater use of simulators, increasing use of companion trainers, shifting high-severity-factor flight events to other aircraft, reducing the aircrew qualifications, and eliminating high-severity-factor training.

Increase Use of Simulators for Training

Approximately one-third of the flying-hour program is devoted to event-based training required for maintaining pilot proficiency. This subsection analyzes the amount of training that can be accomplished in the simulator, the flying-hour savings, and the resulting reduction in EBH accumulation. The resulting EBH reduction will then be applied to current usage rates to determine the effect on the drawdown of the current fleet and the delay allowed in recapitalizing the C-130 fleet.

Depending on the specific event, the training regulation allows some event-based training to be accomplished in the simulator. Generally, USAF regulations allow pilots to complete all pilot-proficiency events, such as takeoffs, landings, and instrument approaches, in the simulator. Regulations also allow pilots to complete half of all required tactical training events, such as tactical approaches and arrivals, night-vision-goggle training, visual low-level flying, and assault landings, in the simulator. Only two events for Mobility Pilot Development must be accomplished in the aircraft: (1) left-seat landing and (2) left-seat tactical sortie.

AMC already assumes that some training requirements will be met using the simulator. Currently, after their initial training, C-130 pilots attend simulator refresher training once a year, during which they receive instruction and practical experience in crew resource management and emergency procedures. During this training, pilots also accomplish a number of required events in the training tables, and AMC assumes that 15 percent of the annual basic proficiency events will be accomplished there. ANG and AFRC do not allow pilots to use the annual simulator training as credit for any of these requirements.

There are advocates in the airlift community for increased simulation who use the practices of major commercial airlines as an example of why simulators could be used for more training in the Air Force. Airlines do all their training in simulators, and aircraft are flown only on revenue-generating flights. Others argue that military flying training in the aircraft needs to be preserved because new Air Force pilots have less experience than new airline pilots. New airline pilots have a minimum of 3,000 hours of experience in other aircraft, while new C-130 pilots generally have only 200 to 300 hours.

In light of these differences, we assessed two levels of increased simulation: (1) accomplishing only basic proficiency training in the simulator and (2) accomplishing both basic proficiency and half of all tactical training in the simulator.[3] We used the AMC flying-hour model to calculate required flying and savings. Accounting for the additional aircraft of the Pacific Air Forces, U.S. Air Forces Europe, and the ARC, implementing the respective options could save approximately 18,000 and 35,000 flight hours annually.

Determining how this flying-hour savings would affect fleet life required converting the reduction in flying into a corresponding savings in EBH. We converted flying hours to EBH by calculating average severity factors for the types of flying that would be moved to the simulator. The training events fell into two broad categories: training that included low-level, high-speed flying and training that did not. The matched sorties from the two databases described earlier (GDSS and AIRCAT) were grouped into missions categorized operationally as training, which were then split into those that contained low-level flying and those that did not. Table 4.3 shows these groupings and

Table 4.3
Training Event and Mission Groups for Calculating Average Severity Factors

| GDSS Mission Class | AIRCAT | | Training Table Events |
	Mission Group	Average Severity	
Training	Training	2.33	Takeoff and landing
			Instrument approach
			Assault takeoff and landing
			Tactical departure and arrival
			Upgrade training
			Penetration and descent
	Low level	3.88	Low-level day and night
			Station-keeping equipment for instrument meteorological conditions and the adverse weather aerial delivery system
			Airdrop

---

[3] Neither of our levels of increased simulation exceeded C-130 aircrew training regulations. The higher level that includes both the basic proficiency and half of the tactical training is the maximum allowed by the regulation.

severity factors. Applying these average severity factors yields an EBH savings of 45,000 and 98,000 for the respective simulator options.

Figure 4.3 shows this calculation graphically, where "Training" AIRCAT mission group hours are multiplied by an average severity factor of 2.33 and the low-level flight hours are multiplied by a severity factor of 3.88. This results in the number of annual EBH savings hours for Basic only and Basic plus half tactical shown in the figure. Figure 4.4 shows that, if this policy option could be implemented right away, the flying-hour savings would delay the need to recapitalize the C-130 fleet by only one to two years.

Figure 4.5 shows how additional simulators would be needed to provide sufficient capacity to satisfy training events. The horizontal bar at the top of the figure represents required hours—the dark portion for 100 percent proficiency and the lighter portion for 50 percent of tactical. The scale at the bottom of the figure shows the number of simulator hours, and the middle portion of the figure, labeled "additional capacity," shows when additional simulators are needed. For example, achieving 100 percent of specified proficiency hours in the simulator

Figure 4.3
Equivalent Baseline Hour Savings from Additional Simulation

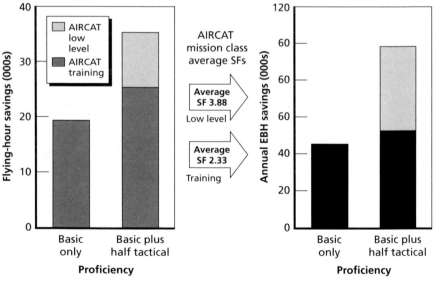

Figure 4.4
How Increasing Use of Simulators Affects the Need to Recapitalize the C-130 Fleet

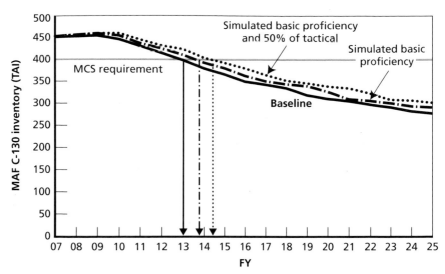

RAND *MG818-4.4*

Figure 4.5
Additional Simulator Capacity Required to Meet Training Events for Simulator Options

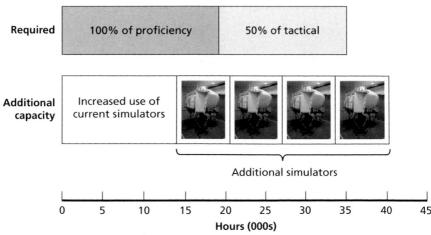

RAND *MG818-4.5*

requires 18,000 hours (the right end of the dark portion of the bar lines up with 18,000 hours on the scale). Meeting this capacity requires the addition of one new simulator. Satisfying 50 percent of tactical hours (an additional 17,000 hours) requires three more simulators.

With its current inventory of simulators, the Air Force does not have enough training capacity to realize the flying-hour savings just described. We assumed that the training hours shifted to the simulator would be in addition to those already programmed for upgrade and refresher training. Currently, Air Force C-130 simulators operate 240 to 330 days per year, 18 hours per day, depending on the location. Increasing simulator operation to 365 days a year and 18 hours per day would make an additional 15,000 hours of simulator time available.[4] Increasing operating hours to this level would still not provide the number of hours required even to perform all the basic proficiency training in the simulator. Figure 4.5 shows that one additional simulator would be required to conduct the basic proficiency training and that four additional simulators would be required to conduct both the basic proficiency training and one-half of the tactical training.

Along with the capacity challenges, two practical challenges for the Air Force to conduct so much more simulator training stand out: (1) lack of a common cockpit configuration and (2) travel requirements for C-130 units that do not have a collocated simulator. At present, there is only one simulator for each of the H1, H2, and H3 models, while there are seven C-130E simulators. Table 4.4 summarizes the number of aircraft, simulators, simulator locations, and aircraft operating locations for all C-130 models. Each of the three C-130H simulators supports a much larger population of aircraft than does each C-130E simulator.

In January 2007, the Air Force initiated the C-130 Avionics Modernization Program, which, when complete, will mitigate this lack of commonality in the airframes and simulators. C-130 units are located at 40 operating locations around the world; however, currently, only seven of these locations have simulators. (One of the simulator loca-

---

[4]   As a point of comparison, commercial airline simulators at a major U.S. airline operate 22 hours a day, every day of the year.

Table 4.4
C-130E/H MDS Aircraft, Simulators, and Locations
(no.)

MDS	Aircraft		Simulators	
	Total	Operating Locations	Total	Locations
C-130E	129	13	7	4
C-130H1	47	3	1	1
C-130H2	149	17	1	1
C-130H3	80	7	1	1

tions, McChord, does not even have an associated C-130 unit.) Because many of the training events are required monthly, a unit that did not have a simulator on its own base would face significant travel. This travel could burden some ARC units.

AMC Aircrew Operations and Training Division (AMC/A3T) currently has plans to build additional simulators to help alleviate the capacity limit and partially mitigate the travel requirements. Figure 4.6 presents the locations of all the simulators and operational units.

Increase Use of Companion Trainers

A CTA program could potentially reduce the rate of accumulation of fatigue damage on C-130s by allowing less-experienced pilots to gain airmanship and the hours required to upgrade to aircraft commander by flying an alternative aircraft. The program would provide seasoning analogous to the Accelerated Copilot Enrichment program that the Air Force had in the 1980s and 1990s.

The number of flying hours required to upgrade to aircraft commander depends on whether a pilot enters the C-130 crew force directly from UPT or after having had experience in another aircraft. Pilots who enter immediately after UPT (and have about 200 hours of flying time) require 700 hours of C-130 time (for a total of 900 hours) to become an aircraft commander. Pilots who enter after experience in another aircraft are called prior-qualified pilots (PQPs), and the amount of C-130 time they need to become an aircraft commander depends

Figure 4.6
Locations of Operational Units and Simulators

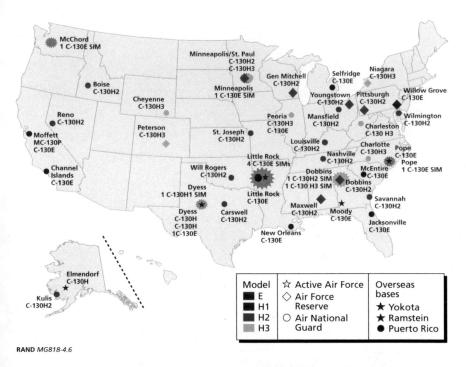

RAND *MG818-4.6*

on the number of hours in the prior aircraft. For example, a PQP with 1,000 hours in another aircraft would require only 100 hours of C-130 time to become an aircraft commander.[5]

For our proposed CTA program, copilots would still fly in the C-130 to maintain their currency and conduct required mission training. However, instead of making up the difference between training-required flying and the total flying required to become an aircraft commander with user-funded hours, these hours would be flown in the CTA. To help make up for the decreased flying experience in the C-130, we increased the total flying hours required for UPT inputs to become an aircraft commander to 1,100 hours—that is, the same total number of flying hours required for PQP inputs. This is illustrated in Figure 4.7. The left bar in the figure shows that PQP inputs

---

[5]   AFI 11-2C-130V1, Table 5.1, p. 44.

**Figure 4.7
Minimum Flight Hours and Proposed Companion Trainer Program for
Aircraft Commander Upgrade**

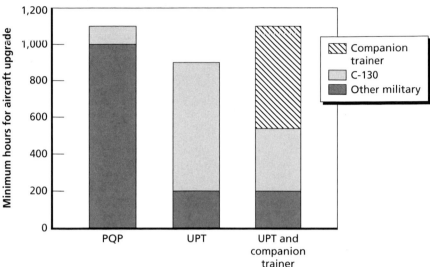

RAND *MG818-4.7*

with 1,000 hours of prior flying time require 100 hours in the C-130 to become an aircraft commander. The second bar shows that, currently, UPT graduates start C-130 training with about 200 hours of flying time and need 700 hours of C-130 time before upgrading to aircraft commander. The third bar shows that, with the CTA, they will upgrade to aircraft commander with 1,100 total hours (just like PQP inputs) but with 340 hours in the C-130 and 560 hours in the CTA.

The potential flying-hour and EBH savings from implementing the CTA option are 30,000 and 52,000, respectively. The flying-hour savings in this option would come primarily from user-funded hours, which, on average, have a low severity.

Figure 4.8 shows that immediate implementation of a CTA program offering the EBH savings noted above would delay the need to recapitalize the C-130 air mobility fleet by less than a year. The high EBH accumulations of so many C-130s today are one of the principal reasons this option offers such limited leverage.

Figure 4.8
How a Companion Trainer Program Affects the Need to Recapitalize the C-130 Fleet

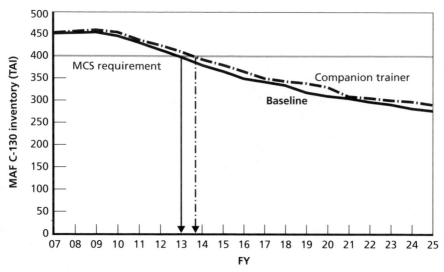

Implementing a CTA program today has some challenges. These include increasing required copilot flying, the need to fund a large fleet and flying-hour program for CTA, possible loss of revenue to the Air Force, and loss of net C-130 experience to the crew force.

The number of required monthly flight hours for each copilot would increase, from 25 to over 32, to allow them to acquire 900 hours before upgrading to aircraft commander. As described earlier, 900 hours would be provided as an equivalent number of flight hours at aircraft commander upgrade to pilots coming from UPT to pilots coming to the C-130 from a post-UPT assignment.

A CTA program like the one described would be quite large, having an annual flying-hour program of approximately 47,000 flight hours and a fleet of approximately 50 aircraft.[6]

---

[6] The flying-hour program and required aircraft for the CTA assumes 114 active-duty primary aircraft inventory C-130 aircraft. The fleet has a crew ratio of 2.0, and each crew has two pilots, resulting in 456 pilots. AMC assumes an experience ratio of 43 percent, meaning that 196 pilots need to be upgraded at any time. The training program lasts 28 months and

Substituting CTA hours for user-funded hours in the flying-hour program could result in a revenue loss to the Air Force. While this is not as big a concern for DoD at large, this could be particularly challenging for the Air Force in funding a CTA program. The hours that would be moved to the CTA are hours that are typically funded through user fees from agencies outside the Air Force. If these flying hours were replaced by hours funded by the Air Force, the result would be a significant net loss to the Air Force budget.

Two potential areas of concern associated with a CTA program are (1) a net loss in C-130 experience in entering aircrews and (2) the challenges of relatively inexperienced pilots maintaining qualification in two different airframes. Under the assumptions for our CTA analysis, when UPT inputs were able to upgrade, they would have about half the C-130 hours of new aircraft commanders under the current system. While saving flight hours, this reduction in C-130 experience would also decrease the overall experience level of the active-duty C-130 crew force because over 90 percent of all pilots entering the C-130 come from UPT. Another concern about the proposed CTA program is that, while PQP pilots are allowed by regulation to upgrade with only 100 C-130 hours, these pilots are in fact typically managed by unit leadership, especially when they are first upgraded. By *managed*, we mean they are paired with more-experienced copilots, flight engineers, navigators, and loadmasters. Also, these pilots typically fly less-complex missions.

Despite the challenges, a CTA program could make good sense, especially if the Air Force was having difficulty finding enough user-funded hours to meet its training needs. A shortage of this nature occurred in the late 1990s. However, currently, with C-130 operations and commitments around the world, especially in the Middle East, there is no shortage of user-funded hours to provide the needed seasoning hours. Not only is there not a shortage, there would be little time

---

Figure 4.7 shows that 560 CTA hours are required. This yields a flying-hour program of just over 47,000 hours for the CTA program (560 × 114 × 2 × 2 × 0.43 × 12 ÷ 28). Compared to other small, nontactical aircraft in the Air Force inventory, a large number of flying hours per year is the assumption for the CTA, which is favorable for the cost analysis, which still shows the CTA program yielding a negative NPV.

available for the inexperienced pilots to even fly the CTA. Figure 4.9 makes the magnitude of the commitment to contingency operations since September 11, 2001, readily apparent.

Shift High-Severity-Factor Operational Missions to Other Aircraft

Contingency and channel missions together make up about 33 percent of the total flight hours and about 27 percent of the total EBH. The average severity factor is less than one, indicating that, overall, these missions are fairly benign. However, some of these missions have higher severity factors than others. We considered shifting channel and contingency operational missions with a severity factor greater than two and payloads greater than 6 tons to the C-17. These missions account for about 4 percent of the total flight hours and 10 percent of the total EBH. Figure 4.10 presents the percentage of flying hours and EBH accumulation for these missions. Figure 4.11 shows that recapitalization can only be delayed by about a year using this approach. Using the C-17 to haul C-130 loads has significant drawbacks if payload

Figure 4.9
Air Mobility Command Active-Duty C-130 Flying-Hour Breakdown,
FYs 1998–2006

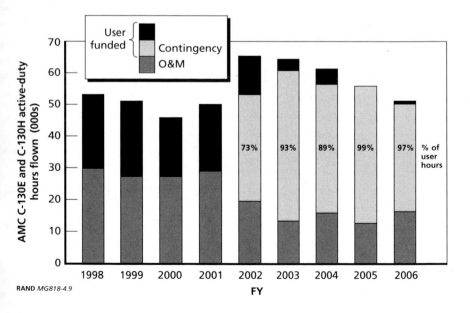

Figure 4.10
Share of Hours for Contingency and Channel Missions

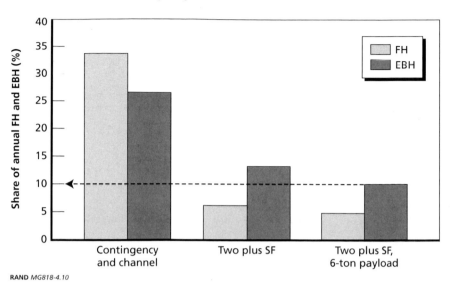

RAND *MG818-4.10*

Figure 4.11
How Shifting Some Missions to C-17 Affects the Need to Recapitalize the C-130 Fleet

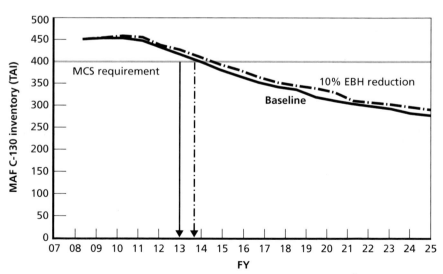

RAND *MG818-4.11*

cannot be consolidated. The C-17 is a much larger platform and the level of consolidation will affect the overall viability of this approach. We assumed that 15 percent fewer C-17 sorties would be required to account for this payload consolidation. This is discussed in greater detail in Chapter Seven. This, of course, could have an adverse impact on the C-17 life and result in the need to recapitalize this fleet sooner than expected.

Reduce Crew Qualifications and High-Severity-Factor Training

The analysis of the training events in the subsection on simulators showed that the events required to maintain aircrew formation-airdrop qualification were, collectively, the high-severity-factor events. For this analysis, we examined the effect of eliminating the formation-airdrop qualification for one-half of the aircrews. Currently, the entire C-130 crew force is formation-airdrop qualified. Using the same methodology discussed in the simulator subsection, we determined a potential savings of 11,000 flight hours and 40,000 EBH by maintaining airdrop qualification for only one-half of the crew force.

Although these missions have a high severity factor—nearly 4— the number of flight hours and resulting EBH savings are relatively small. As a result, the potential to delay recapitalization using this option is only about six months.

If this option were implemented, the consequences for validated Army paratroop requirements would need to be examined. Additionally, this option might increase scheduling complexity when airdrop operations are required operationally.

Options Eliminated During Initial Screening Process

In this section, we present options that were eliminated from consideration in the initial screening process. In this section, we describe the options and present the rationale for elimination.

Rotate Aircraft Among Components

We examined a case in which we allowed the Air Force to trade active aircraft with little service life remaining for ARC aircraft that had more service life remaining. Swapping high-use aircraft for low-use aircraft could potentially extend the useful service life of the current fleet by reducing the usage rate of the aircraft with the least life remaining. This would not change the number of aircraft retired over the long run because neither the total fleet usage nor the amount of service life on the aircraft would change.

This option is illustrated in Figure 4.12, which shows all the current C-130Es and C-130Hs. Only aircraft that were being used at rates of over 1,000 EBH per year were considered for the exchange.[7] The ellipse and rectangle on this chart give the reader a graphical sense of which aircraft are being rotated in this option. Note that most of the aircraft with the least amount of service life remaining are in the active

Figure 4.12
Accumulation of Equivalent Baseline Hours Relative to Remaining
Equivalent Baseline Hours

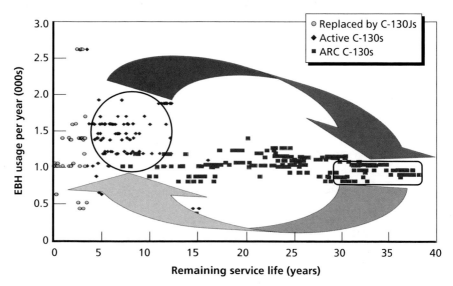

NOTE: Aircraft with least service life remaining and high usage are swapped for those with most service life remaining and lowest usage.
RAND *MG818-4.12*

---

[7] Including aircraft that were being used at lower rates could have increased the use of these aircraft. This, in turn, would have moved up their retirement dates, which was just the opposite of what we were trying to achieve.

component. Note also that ARC aircraft tend to use up less EBH each year. Most ARC C-130s are used at 800 to 1,000 EBH per year, while many C-130s in the active component are used at rates well in excess of 1,000 EBH per year.

Gray dots indicate aircraft that are scheduled to be replaced by C-130Js over the next few years. These aircraft were not candidates for the swap. Since they are to be retired anyway, swapping them would not extend the service life of the fleet and thus would not delay the need for SLEPs or new buys. Aircraft were selected for the swap based on their remaining service life. The 100 aircraft with the least amount of service life remaining (regardless of their assigned major command) were exchanged for the 100 aircraft with the greatest amount of service life left. The usage profiles remained identical; only the aircraft were exchanged.

For example, consider two aircraft, one in the active component and one in the ARC. Assume that the active aircraft had 4,000 EBH remaining and was being used at 1,000 EBH per year, while the ARC aircraft had 10,000 EBH remaining and was being used at 500 EBH per year. The active aircraft would have four years of service life remaining, while the ARC aircraft would have 20. Swapping these aircraft would increase the service life of the active aircraft from four to eight years and reduce the service life of the ARC aircraft from 20 years to 10 years.

The projected C-130 inventories for both the baseline and the 100-aircraft-swap case are illustrated in Figure 4.13. While the swap delays the retirement for half of those swapped (by subjecting them to a lower EBH usage rate for their remaining life), it moves up the retirement date for the other half. The net effect is a delay of year or two for approximately a dozen aircraft. The point at which the C-130 inventory falls below the MCS requirement is effectively the same. Thus, this option does not delay the need for SLEPs or new aircraft.

Even if it were possible to delay the need for SLEPs or new aircraft buys through the use of aircraft swaps, it may not be desirable to do so. The Air Force and ARC operate five mobility C-130 models, each requiring a separate pilot qualification. Swapping 100 aircraft would require hundreds of pilots to go through additional training to become

Figure 4.13
How an Active-Reserve Aircraft Swap Affects the Recapitalization Timeline

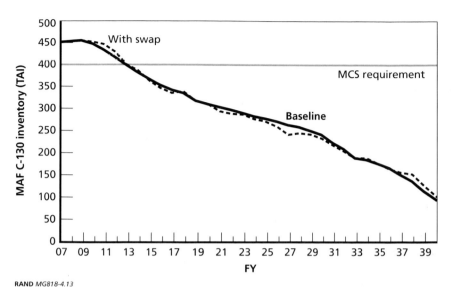

qualified on the new model. Although parochial concerns must not drive the decision, the AFRC and ANG units may fight hard to keep their relatively new, highly capable C-130Hs rather than trade them for old C-130Es, with their much more-limited performance. These concerns, however, are secondary; Figure 4.13 shows that the swap of 100 aircraft is not an effective option to delay the need to recapitalize the C-130 fleet.

Increase Experience Mix

This option considered the possibility of staffing C-130 units with more-experienced pilots to reduce the seasoning burden significantly and the amount of event-based training. Currently, over 90 percent of pilots entering the C-130 crew force come directly from UPT. These pilots require AMC to provide seasoning and more currency training because they have less flying experience. For this option to work, inputs to the C-130 crew force would need to have had at least one previous flying assignment. This concept is not without precedent; the executive transport wing at Andrews AFB is staffed with highly experienced pilots, as was the C-5 Galaxy during its initial operation.

While this option would reduce the training burden on the C-130 fleet, other mission design series (MDS) communities would have to shoulder the burden and provide experience to more pilots directly from UPT. Conducting seasoning in other aircraft, such as the C-17 or the KC-10, may be more costly on average.

In addition to the shift in training burden, there is a concern about officer professional development. In a crew force staffed with highly experienced pilots, there would be a limited number of Air Force positions available commensurate with the officers' experience levels. Pilots flying C-130s may lack the command opportunities of their counterpart in flying other MDSs. Having experienced captains and majors serving in positions traditionally filled by lieutenants may negatively affect the promotion potential of these pilots.

As a result of these significant negative consequences for the force, we dropped this option in the initial screening process.

Change Active-Reserve Mix

We looked at both increasing and decreasing the fraction of active-duty C-130s. Increasing the number of active-duty C-130 forces would effectively make more C-130s available for ongoing deployments and commitments. Further, this increased availability could potentially reduce the accelerated wear on the active-duty aircraft and crews during extended deployed operations. However, this "leveling" of the EBH accumulation would not have much effect on the time frame of recapitalization, since this is similar to swapping active and ARC aircraft. The other major concern with this option is that the increase in active-duty aircraft and crews would create an additional training burden. Over the long term, this would increase the EBH accumulation on the aircraft.

Doing the converse, reducing the fraction of active-duty units, could reduce overall EBH accumulation on the fleet because less training would be required because of the higher experience mix. This, however, would have significant detrimental consequences for the ability of the total force to support a long war. As discussed in Chapter Seven,

ARC units are less available for overseas commitments than active-duty units. As a result, the remaining active-duty units would have to operate at a lower dwell-to-deploy ratio, which could have significant negative consequences for the quality of life and retention of aircrews.

Although there might be other good reasons to consider changes in the active-reserve mix, such changes appear to have enough negative implications to reject them as a means to delay the need to recapitalize the C-130 fleet.

Add ARC Associate Units to Active Squadrons

The concept behind this option is to provide the required crews per tail by adding more-experienced ARC pilots at active-duty locations in the form of associate reserve units. This increase in experienced pilots would reduce the amount of training required and would lower personnel costs because a vast majority of the crews would be part-time reservists or guardsmen.

However, accomplishing this option and realizing the potential savings would require a complementary reduction in the active-duty crew ratio. This, in effect, would place more of the C-130 capability in the reserve forces, limiting the number of crews available for the deployment obligations of a long war.

The marginal increase in flying savings, the extended time required to stand up additional guard and reserve units, and the reduction in on-hand capability eliminated this option from further consideration.

Increase Squadron Size

This option considered combining active-duty squadrons, thereby reducing the number of command and overhead pilot positions and potentially saving flying hours. An active-duty squadron has only two overhead flying positions, the squadron commander and the director of operations. Because of the limited number of active-duty locations with multiple squadrons, only approximately ten positions could be saved. The flying-hour savings for these ten positions is also less than average because these positions are typically filled with highly experienced pilots and have minimal continuation and upgrade training requirements.

The increased workload this would mean for the staff pilots in the larger squadrons, along with the minimal flying-hour savings, limited the potential of this option to delay the need to recapitalize the C-130. We therefore eliminated it from further consideration.

Place Flight Restrictions on Specific Aircraft

This option considered placing additional restrictions on the aircraft before they reach 38,000 EBH; these restrictions would attempt to limit maneuvers that increase EBH. As has been shown throughout this chapter, reducing the EBH accumulation on each aircraft by even a sizable fraction delays the need to recapitalize by only a few years.

This option, however, has significant negative implications for retaining mission-ready crews. This analysis eliminated this option from further consideration because limiting the maneuvers reduced the overall training effectiveness and capability of the fleet.

Observations

None of the options for reducing the rate of EBH accumulation significantly delayed the need to recapitalize the C-130 fleet. Moreover, many of these options had negative implications and/or barriers that could limit implementation. We did, however, assess whether selected options offered cost savings, despite their inability to significantly close the capability gap identified in the FNA. These results are reported in Chapter Eight.

CHAPTER FIVE

Delaying C-130 Recapitalization by Increasing the Supply of EBH

This chapter examines a set of options (both materiel and nonmateriel solutions) characterized as increasing the supply of EBH. Table 5.1 presents the options identified during the first IPT meeting. The materiel solutions—SLEPs and new aircraft buys—receive only limited attention here because these will be addressed in detail in the UIAFMA.

Table 5.1
Options to Increase Supply of Equivalent Baseline Hours

Options	Estimated Potential Impact	Other Implications
Most promising		
SLEP/repair the aircraft	High	Risks associated with aging aircraft
Buy additional aircraft	High	Additional capability Greater flexibility Reduced risk
Accept greater risk	High	Greater risk of catastrophic failure
Dropped in the screening process		
Develop better diagnostic tools	Moderate	Reduced uncertainty

Key:
Green, *few* or *none*
Yellow, *moderate*
Red, *significant*

Options Analyzed

SLEP or Repair the Aircraft

This section discusses three approaches to structural mitigation: prevent damage, inspect and repair, and refurbish or replace. All these can extend the life of an aircraft, and all, to some extent, have a materiel component:

- **Prevent Damage.** One established means of preventing damage involves inducing a zone of residual compressive stresses around and through a hole. Typically extending radically, at least one radius around the hole, the compressive stresses from "cold working" retard the initiation of fatigue cracks and slow their propagation if any occur. This technique can be used when inserting fasteners during initial production to improve the fatigue life and durability and damage tolerance of new structures or during repairs of structures that have suffered fatigue damage. Thus far, this technique has been applied only to individual structural details on C-130s.[1] The efficacy of applying such a technique to an entire wing box or other major structural component remains to be demonstrated.
- **Inspect and Repair.** Inspections, such as TCTO 1908, are an effective means of allowing aircraft to reach the assessed CWB service-life limit of 45,000 EBH without flight restrictions. However, current inspection approaches are not a reliable means of allowing flight beyond 45,000 EBH because of poor probabilities of inspection and detection of fatigue damage and of the presence of multisite and multiple types of damage on high-EBH aircraft. There is a limit to the number of times this process can be conducted; oversizing holes is often required, which limits repeatability. TCTO 1908, for example, can be conducted only once.
- **Refurbish or Replace.** This technique can range from maximizing the use of existing components through refurbishment

---

[1] Len Reid, Fatigue Technology Inc., "Aging Aircraft Repair Strategies Utilizing Cold Expansion Technology," presented at the 2005 USAF Aircraft Structural Integrity Program Conference, Memphis, Tenn., November 29–December 1, 2005.

to replacing a CWB completely. As discussed in Chapter Two, retrofitting new CWBs is an established technique for increasing the structural service life of C-130s.[2] The cost of this technique will depend on the degree of refurbishment or replacement, but this is generally among the more costly of structural remediation options.

Table 5.2 shows Air Force estimates of the time and resources to accomplish three wing-fatigue mitigations: (1) TCTO 1908 inspection and repair, (2) rainbow fitting replacement, and (3) CWB replacement. Each action involves considerable aircraft downtime. The CWB replacement costs the most, takes the most time to accomplish, and offers the most additional CWB EBH. Any follow-on analyses of mate-

Table 5.2
Air Force Estimates of Time and Resources to Accomplish Center Wing
Fatigue Mitigation Actions

Mitigation Action	Preferred Timing (EBH)[a]	Cost ($M)	Time to Accomplish (months)	Effect of Mitigation Action
TCTO 1908 inspection and repair of lower wing surface				
Inspect	38,000	0.450	4	Unrestricted flight operations to 45,000 EBH
Repair		0.250		
Rainbow fitting replacement				
Programmed depot maintenance (PDM)	24,000	0.375	TBD	24,000 EBH for replacement fitting
Unscheduled depot-level maintenance		0.515	2	
Center wing replacement	45,000	9.000	6	45,000 EBH for replacement wing[b]

SOURCES: Fraley, 2005; Fraley and Christiansen, 2006.

[a] No later than specified.

[b] Other factors will likely prevent realization of full life enhancement.

---

[2] Depending on the state of an airplane, aging of other structural components or functional systems could limit full realization of the life extension from a CWB replacement (see Chapter Two and later in this chapter).

riel options will need to examine critically the estimated costs of these mitigation and other structural mitigation options, as well as the costs to keep other aging C-130 systems viable.

A materiel solution, such as CWB replacement, could add significant life to C-130s (see the shaded area in Figure 5.1). The case in the figure assumes that either all E and H models receive replacement CWBs or just H2s and H3s. The replacement yields 15 to 20 years of additional life before other structural constraints come into play. As noted earlier, however, the Air Force would undoubtedly need to address problems that other aging aircraft systems already pose to realize the additional life.[3] The UIAFMA delves further into this subject.[4]

Figure 5.1
A Materiel Solution (Center-Wing Replacement) Can Add Significant Life

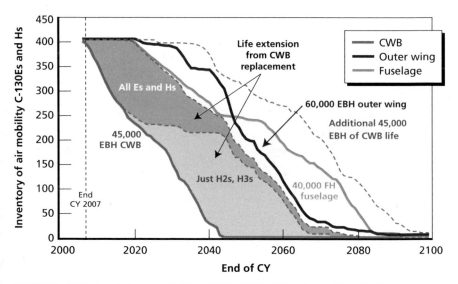

SOURCES: CWB sheet, January 2007; AIRCAT, 2007; *ASIP Master Plan*, 1995; LMAC, 2006; S. F. Ramey and J. C. Diederich, "Operational Usage Evaluation and Service Life Assessment," presented at the 2006 Hercules Operators Conference, Atlanta, Ga., October 2006.
RAND *MG818-5.1*

---

[3] The dashed line to the far right in Figure 5.1 shows the additional life gained from a CWB replacement, ignoring other life-limiting structural constraints.

[4] Kennedy et al., 2010.

Structural mitigation actions can address specific aging issues, but aircraft are still subject to chronological aging processes, most notably corrosion. With SLEPs, fleets get much older, increasing the possibility of numerous age-related problems. Figure 5.2 illustrates the chronological aging of C-130E and H aircraft with and without a CWB replacement. The solid line shows the average age of the C-130E and H fleet if the Air Force retires C-130s at 45,000 EBH. In this case, the average age would ultimately reach 50 years by retirement of the last airplane.

Aging is much more dramatic if aircraft undergo a SLEP. The lower and upper dashed curves define the average and maximum age respectively, assuming all C-130Es and Hs undergo a CWB replacement and are retired when they reach outer-wing or fuselage service-life limits. In this case, by the time 200 aircraft remain, the average age would be about 60 years, and the oldest plane would be about 85 years

Figure 5.2
With SLEPs, the Fleet Gets Much Older

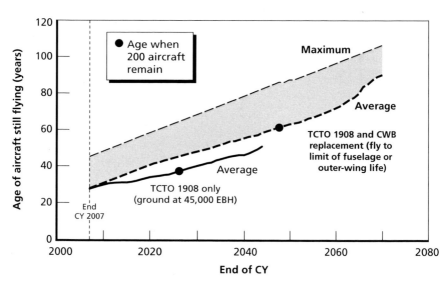

SOURCES: CWB sheet, January 2007; AIRCAT, 2007; *ASIP Master Plan*, 1995; LMAC, 2006; Ramey and Diederich, 2006.
NOTE: Air mobility C-130Es and Hs.
RAND MG818-5.2

old.[5] Continued aging beyond that point would put the C-130 fleet into largely uncharted territory in terms of fleet aging.[6] Although other USAF fleets may be of similar age (e.g., B-52 and KC-135), the various MDSs will have different problems (e.g., fatigue, corrosion) associated with the different flight characteristics they have experienced during operations.

Buy Additional Aircraft

Recapitalization of the fleet with a new aircraft is also a materiel solution and is therefore analyzed in the UIAFMA.[7] Determining which aircraft type to procure is the role of an AoA. The different aircraft types that could be considered include aircraft currently in production or planned to be in production for the USAF (the C-130J-30, the C-27J, and the C-17A and such other aircraft as the European Aeronautic Defence and Space Company A400M).[8]

This analysis, however, focuses on nonmateriel solutions. As a baseline for comparison of costs against those of the nonmateriel options, we made a rudimentary cost estimate for a new aircraft procurement using the C-130J. We then used this baseline to determine which other potential options are cost-effective means of delaying the need to recapitalize the fleet. These comparisons are shown in Chapter Eight.

Accept Greater Risk

Another option is to fly beyond 45,000 EBH and accept a higher risk of flight failure. The rapid increase in single-flight probability of failure for high-EBH aircraft illustrated in Chapter Two translates into a poor reward (extra EBH) to risk (flight failure) ratio. Figure 5.3 illustrates the consequences of flying beyond 45,000 EBH with heightened risks.

---

[5]   No assertion about the practicality of operating such an old fleet is intended.

[6]   Age curves in Figure 5.2 for the CWB replacement case are arbitrarily truncated at 2070. A couple of dozen C-130s would still not have reached structural life thresholds by that time (see Figure 5.1).

[7]   Kennedy et al., 2010.

[8]   As of this writing, first flight of the A400M is expected in 2010.

Figure 5.3
Risk-to-Reward Trade-Off from Flying Beyond 45,000 Equivalent Baseline Hours with More Risk Is Poor

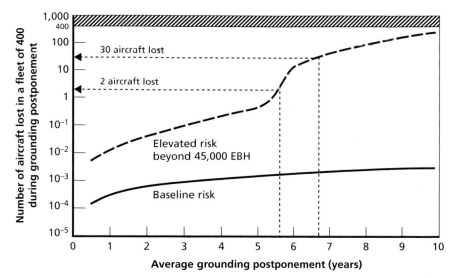

SOURCES: Christiansen, 2006; CWB sheet, January 2007.
NOTE: Potential losses due to structural failures flying when recent risk curve beyond 45,000 EBH.
RAND MG818-5.3

The Air Force would face greater risk of aircraft losses if it chose to fly C-130s beyond 45,000 EBH because the single-flight probability of failure increases rapidly beyond the current grounding threshold. The illustration in Figure 5.3 assumes that the Air Force flies each plane in a 400-aircraft fleet beyond the grounding threshold by the number of years shown on the x-axis. The dashed curve shows the cumulative number of aircraft losses over time as the single-flight probability of failure shown in Figure 2.10 increases with EBH.[9] For comparison, the solid line shows the minimal expected losses if there were some way to keep the probability of failure within MIL-STD-1530C standards.

---

[9] These calculations assume accomplishment of one TCTO 1908 inspection to control flight risks prior to 45,000 EBH. Generalized cracking, poor probability of detection (POD) and probability of inspection (POI), inspection repeatability constraints, and frequent and lengthy downtime to complete inspections would limit the practicality of continued inspections.

Figure 5.3 does not represent USAF policy but is based on our statistical calculations using the expected increase in risk and flying-hour profiles. Note that flying a little more than five years beyond the current grounding threshold suggests the loss of two extra aircraft.[10] Within another year, aircraft losses could rise to double-digit levels. The amount of fleet life gained from flying beyond the fatigue threshold does not appear to justify the significant risk of aircraft losses.[11]

Options Eliminated During Initial Screening Process

Develop Better Diagnostic Tools

The last option listed in Table 5.1, the development of better tools for diagnosing structural health, would not immediately delay the need to recapitalize but does have the potential to improve estimation of the fatigue damage each aircraft is accumulating. Such tools could include instrumentation to measure operational loads and to capture flight parameters, flight events, and weights automatically, in flight. Other methods of acquiring more knowledge could include selective structural teardowns and periodic sub- and full-scale fatigue tests. Collectively, better tracking of individual aircraft, coupled with more-extensive testing, would significantly enhance understanding of the health of the fleet.

An emerging approach for assessing the health of structures is to continuously monitor their condition using a network of sensors that can be embedded in or attached to an aircraft. These sensors may be part of a load-monitoring system or could use a variety of sensing techniques to detect damage directly and determine its location when it

---

[10] Recall that the loss of two USAF C-130E aircraft because of outer-wing fatigue failures in the 1980s precipitated an outer-wing redesign and retrofit of all C-130Es.

[11] If, instead of consciously deciding to fly beyond the EBH grounding threshold, a crew inadvertently flew beyond the limit because of errors in estimating accumulated EBH, there would also be a rapidly escalating risk of aircraft losses. Even with a small probability of underestimating EBH, the risk of aircraft losses could be significant because of the steep slope of the single-flight-probability-of-failure curve at higher EBHs (see Figure 2.10).

occurs.[12] Structural health monitoring (SHM) permits condition-based maintenance, in which the need for maintenance is driven by a failure or incipient failure condition detected by sensors rather than detected during inspections and maintenance at fixed intervals. In principle, restricting maintenance to only the times human intervention is really needed could reduce the cost of inspections and mitigate at least some of the human-factor problems associated with more-traditional NDIs, while reducing downtime for aircraft operators.

SHM can employ a variety of sensing technologies. In fact, multiple sensing technologies would probably be required for any full aircraft implementation to meet all the damage-detection requirements for various combinations of materials, structural arrangements, and flight loadings, as well as damage-monitoring requirements. Among the technologies are acousto-ultrasonics, comparative vacuum monitoring,[13] acoustic emission, microwave sensors, imaging ultrasonics, and foil eddy-current sensors. Each of these technologies has special fields of application, and, in that sense, they complement one another. Some are designed to monitor hot spots, known locations where fatigue cracks could develop. Others are designed for more-global surveys of potential damage. Some can be conveniently incorporated only during manufacturing, while others can be fitted to existing planes.[14]

Major airframe manufacturers characterize SHM technologies' readiness as follows: "proof of concept in real structures under in-service operational conditions is still lacking . . . ."[15] There are some

---

[12] Reportedly, to date, load monitoring systems using strain gauges for example have been used more commonly than damage monitoring systems for in-service aircraft. Holger Speckman and Rudolf Henrich, "Structural Health Monitoring (SHM): Overview of Airbus Activities," presented at the 16th World Conference on NDT, Montreal, Canada, August 30–September 3, 2004.

[13] A paper presented at the 2006 Hercules Operators Conference describes applications of this technology in full-scale fatigue testing and flight trials of helicopters and fixed-wing aircraft. The same paper proposes potential applications for C-130s (Andrew Chilcott, Structural Monitoring Systems, Ltd, "Comparative Vacuum Monitoring CVM™," presented at the 2006 Hercules Operators Conference, Atlanta, Ga., October 2006).

[14] Speckman and Henrich, 2004.

[15] Speckman and Henrich, 2004.

recent indications of progress, however. Boeing has reportedly incorporated some SHM techniques into its NDI standard practices manual for Boeing airframes.[16] Airlines are considering SHM applications with the assistance of the Federal Aviation Administration and Boeing to address specific maintenance requirements.[17] The helicopter industry takes advantage of vibration data trending for predictive maintenance.

Researchers draw clear distinctions between damage diagnosis, which is the act of identifying when something is wrong, and the embryonic state of damage prognosis, which involves assessing the current state of a system and estimating its remaining useful life, or determining that a problem is imminent and an aircraft must land.[18]

SHM's departure from current practice will require significant test and analysis, as well as evolutionary demonstrations to build confidence in the approach. Researchers suggest initially applying SHM to problems with "well-defined damage concerns" in parallel with current system evaluation and maintenance procedures until its reliability and cost-effectiveness can be proven.[19] Several researchers suggest the desirability of deploying SHM sensors on unmanned aerial vehicles before moving to widespread adoption on manned aircraft.[20]

Given the technology's state of readiness, the generalized cracking characteristics of high-EBH C-130s (see Figure 2.8), and the orders of magnitude escalation in risks that can occur when flying beyond 45,000 EBH (see Figure 5.3), relying on SHM to fly beyond the established CWB service-life limit as a near-term option appears problematic. However, since the Air Force will be flying C-130s for decades to come, it is quite possible that, as selective applications of SHM technol-

---

[16] Sandia National Laboratories, "Sensors May Monitor Aircraft for Defects Continuously, Structural Health Monitoring Systems Accepted by Boeing, Validated by Airlines," news release, Albuquerque, N.M., July 18, 2007, p. 2.

[17] Sandia National Laboratories, 2007.

[18] Charles R. Farrar and Nick A. J. Lieven, "Damage Prognosis: The Future of Structural Health Monitoring," Philosophical Transactions of the Royal Society A, December 12, 2006.

[19] Farrar and Lieven, 2006, p. 631.

[20] Speckman and Henrich, 2004; Farrar and Lieven, 2006.

ogies mature, they could, at the very least, improve maintenance costs and aircraft availability for C-130 flight within established service-life limits.

Observations

This chapter presented our evaluations of several options for increasing the supply of EBH. Aircraft repair covers a wide range of possibilities. New aircraft components and replacement of major structural components are the only ways we found to significantly delay the need to recapitalize the fleet. We distinguished between more-modest repairs, which could extend the service life of the aircraft (such as TCTO 1908), and those that require the replacement of major components, which we refer to as SLEPs. SLEPs have costs on the order of new aircraft buys (see Table 5.2 and the cost analysis presented in Chapter Eight) and are therefore left as an option for the UIAFMA.[21] The useful life that can be expected from any SLEP is also affected by the remaining life of components that are not replaced in the SLEP and the technical risks associated with the uncertainty of operating an aircraft that still has a significant portion of its structure and/or major subsystems remaining from the original production.

This chapter considered two nonmateriel options. Neither was found to be a viable option for delaying the need to recapitalize. We found that operating the aircraft beyond 45,000 EBH without major structural modifications has a poor ratio of risk of flight failure to reward (additional EBH). Better diagnostic tools would provide more-accurate information on the health of the fleet. This information would be highly useful at determining the amount of service life available, and we suggest USAF consider implementing the options we have discussed here. But implementing these tools at this stage in the life of the C-130 fleet would not immediately affect the timing of the need to recapitalize.

---

[21] Kennedy et al., 2010.

CHAPTER SIX

Delaying C-130 Recapitalization by Reducing the Number of Aircraft Needed to Meet the Requirement

In this chapter, we examine the potential for delaying the need to recapitalize the C-130 fleet by reducing the number of aircraft needed to meet the MCS requirement. The 12 options we examined are listed in Table 6.1. During the initial screening, two of the options were found to have the potential to reduce the number of C-130s needed without any significant negative implications and were subjected to a detailed assessment.

Options Analyzed

Shift C-17s to Intratheater Role

The first option was shifting C-17s from the intertheater role to the intratheater mission and backfilling the current C-17 mission with Civil Reserve Air Fleet (CRAF) aircraft.[1] This is viable only when

---

[1] In addition to backfilling the intertheater lift mission with CRAF aircraft, we looked at increasing the crew ratio of the C-17 to allow existing aircraft to perform additional missions. This has significant drawbacks. The planned surge utilization rate for the C-17 is already very high (14.5 hours per day), and it is unlikely that the aircraft could be used more intensively. A higher crew ratio would require more training flight hours per year, thus reducing the expected service life of the platform and increasing the future recapitalization costs for the Air Force. Another option is to acquire additional C-17s. This is a materiel option that is analysed in the UIAFMA.

Table 6.1
Options for Meeting the Requirement Using Fewer C-130s

Options	Estimated Potential Impact	Other Implications
Most promising		
Shift some of strategic lift burden to CRAF and some C-17s to theater lift	High	None[a]
Shift more AETC aircraft during peak demand	Low	None
Dropped in the screening		
Shift some of theater lift burden to surface lift	High	Solution options may not be robust
Fly strategic airlift to forward operating locations (FOLs)	Moderate	Solution options may not be robust
Change theater routes	Low	Solution options may not be robust
Increase maximum number of aircraft on the ground (more civil engineering)	Low	Solution options may not be robust
Increase crew ratio	Low	Solution options may not be robust
Use Joint Precision Air Drop System (JPADS)	Low	Longer load times More training and qualification
Increase Army days of supply	Low	May increase need for tails
Pool joint airlift	Low	Not feasible in some cases
Reduce number of aircraft subjected to a change in operational control (CHOPed)	None	None
Improve in-transit visibility	None	None

Key:
Green, *few* or *none*
Yellow, *moderate*
Red, *significant*

[a] The rating for this option reflects our initial screening. Further analysis indicated that this option is unworkable, principally because meeting the MCS requirement with fewer C-130s could leave the Air Force with inadequate force structure for sustained operations (i.e., the Long War requirement), as discussed later in this chapter.

flying between main operating bases (MOBs) in the continental United States (CONUS) and the theater.

Depending on the level to which this option can be implemented, it could offer great potential. The C-17 can carry three times as many pallets as a C-130 and cruises 100 knots faster.[2] Thus, a few C-17s could substitute for a larger number of C-130s. The USAF has used large numbers of C-130s in recent major combat operations: 149 aircraft in Operation Desert Storm and 124 in Operation Iraqi Freedom (OIF).[3] The actual C-17–to–C-130 substitution rate will depend on the mix of missions. Missions with light payloads that are highly time sensitive (casualty evacuations, emergency resupply, etc.) may require one aircraft regardless of whether it is a C-130 or a C-17. In contrast, missions requiring movement of large amounts of personnel and equipment to a limited number of destinations (i.e., transshipment) may allow C-17s to substitute at a 1:3 ratio. To the extent that high cargo volume missions make up a large share of the need for C-130s, the use of C-17s in the theater role could allow the Air Force to meet the MCS requirements with a smaller number of C-130s.

Using CRAF aircraft to replace C-17s in the strategic airlift role could be particularly attractive. Boeing 747 freighters can carry 34 463L pallets, while the C-17 can carry only 18.[4] Boeing 747s are also somewhat faster. Bulk cargo in the form of 463L pallets typically makes up a large share of the requirement for strategic airlift. While these aircraft are not equipped with defensive systems and thus could not be used in even moderate threat environments, they should be very effective in operating from CONUS to MOBs in secure parts of the theater, where much of the cargo is likely to be moved. Military airlifters could then carry the cargo forward. Because CRAF is civilian owned and

---

[2] AMC, *Airlift Mobility Planning Factors*, Scott AFB, Ill.: AMC Regional Plans Branch, AFPAM 10-1403, December 2003, p. 13.

[3] Lewis D. Hill, Doris Cook, and Aron Pinker, *Gulf War Air Power Survey*, Vol. 5, Pt. I: *A Statistical Compendium*, Washington, D.C.: U.S. Government Printing Office, 1993, p. 31; U.S. Central Air Force, Assessment and Analysis Division, *Operation Iraqi Freedom—By the Numbers*, Shaw AFB, S.C., April 30, 2003, p. 7.

[4] AMC, 2003, p. 12.

operated, no additional procurement expenditures would be needed. A greater use of CRAF may not have any additional peacetime costs if CRAF can provide additional airlift capacity.

Shift More Air Education and Training Command Aircraft During Peak Demand

The second option for reducing the number of C-130s needed that did not have significant adverse implications was to temporarily suspend activities in the dedicated training units at Little Rock AFB and Dobbins Air Reserve Base. Temporarily suspending these training activities would free aircraft that could be used in other roles, reducing the number of aircraft needed to meet the MCS requirement. These aircraft would be flown by instructor pilots assigned to the training squadrons. Training would only be suspended during the peak demand period.

Suspending these training activities would limit the production of new aircrews and maintainers, causing a temporary reduction in the supply of people to C-130 units. Presumably, this shortfall could be made up after the cessation of hostilities. The Air Force has a large pool of aircrews and maintainers qualified to operate the C-130. Temporarily shutting down the training units would not affect any of the units participating in operational requirements because they would already have their full complements of assigned personnel. Shutting down training units would also have a temporary effect on personnel, possibly delaying some assignments. While is precedent for such actions, they have not been common.

The MCS C-130 requirement was based on a combination of contingencies, including two nearly simultaneous major combat operations. The United States has not fought major conflicts in two different theaters simultaneously since World War II. Because this combination of conflicts is very rare, certain changes in established procedures would be justified in dealing with it.

Options Eliminated During Initial Screening Process

Ten of the 12 options were dropped in the screening process. Five of these were not considered robust to changes in scenarios. The underlying scenario assumptions for the MCS were chosen as a result of considerable analysis of the scenarios and global movement capabilities (sealift, rail, etc.). While examining these potential changes in the context of the scenarios, we were able to identify potential changes in the underlying assumptions that might reduce the number of C 130s required to carry out the intratheater airlift mission. We were also able to identify changes in assumptions that would increase the number required. As a result, our evaluation is that changing these assumptions to reduce the number of C-130s may work in some situations but not in others. Therefore, this would decrease flexibility of the force, leaving the resulting force structure less robust in terms of warfighting potential. Further, some of these options present trade-offs that are outside the scope of our analysis—for example, trading between airlift, sealift, and surface lift. Four of the ten options dropped during the initial screening were found to have significant negative implications. We determined that, because of the MCS assumptions, two of the ten options had zero potential to reduce the C-130 requirement. That is, the assumptions in MCS affecting the two options were the "perfect" cases, and changing the assumptions in any way would require more C-130s. The next few paragraphs provide the findings of our analysis of each of the options dismissed during the initial screening.

Shift Some of the Theater Lift Burden to Surface Lift

The first option dropped during the initial screening process was to shift some of the theater lift burden from airlift to sealift or ground transportation. This could be very effective in scenarios involving transfer of large amounts of cargo and personnel over water within the theater. This is particularly true in the Western Pacific, where many of the routes are over water. Many of the ships used for theater sealift (landing craft units, logistics supply vessels, etc.) have large payloads and thus may be able replace large numbers of C-130s.[5] However, these ships travel at only 10 to 12 kts, so their ability to replace airlifters

---

[5]   Joint Publication (JP) 4-01.6, *Joint Logistics Over the Shore (JLOTS)*, Appendix B, Washington, D.C.: Joint Chiefs of Staff, August 2005, p. 2.

would be a function of the time sensitivity of the deployment.[6] Deployments with relatively large arrival windows (i.e., a few days) would be much better suited to the use of theater sealift than those with very narrow windows (i.e., a few hours) because sealift would not be able to meet the required timelines.

While sealift can be very useful in scenarios involving movement of large amounts of cargo and personnel over water, it obviously cannot be used for land routes. It is thus an imperfect substitute for airlift, which can be used in any theater with adequate basing and an acceptable threat environment. While a force composed of some airlift and some theater sealift may meet the mission needs in some scenarios, the same mix may not meet the mission needs in others, such as those taking place in a land-locked theater. Such a force would thus not be robust in the face of changes in the scenario mix. Further, it might not be able to meet the need for time-sensitive deliveries. As a consequence, this option was dropped.

Similarly, ground transportation (i.e., rail or truck) can be effective in theaters where cargo and personnel are largely transported over land and where the infrastructure is adequate. It would obviously be much less effective in theaters with large barriers to land transport (water, mountains, threat troop concentrations, etc.) or where the transportation infrastructure was not well developed. Presumably, MCS considered these factors when deciding to transport cargo and personnel by theater airlift. Because the ground transportation options are highly theater dependent, options to reduce the need for C-130s through greater use of ground transportation options would not be robust to changes in scenarios, so this option was dropped.

Fly Strategic Airlift to Forward Operating Locations
We also examined flying strategic airlifters to FOLs. Theater airlifters have traditionally been used as part of a hub-and-spoke system. Cargo is shipped (via sea and air) to MOBs, then placed on theater airlifters that carry it forward to FOLs.

---

[6] JP 4-01.6, App. B, p. 2.

This has changed somewhat with the introduction of the C-17 and its short-field capability. C-17s were flown into FOLs extensively in Afghanistan and Iraq in Operation Enduring Freedom (OEF) and OIF. Flying strategic airlifters from MOBs outside the theater directly to FOLs would replace two legs (out-of-theater MOB to in-theater MOB to FOL) with a single leg (out-of-theater MOB to FOL). In cutting out the intermediate stop, the C-17 could eliminate the need for the C-130 sortie and thus reduce the need for C-130s.

This option has many potential negatives and is not robust across potential scenarios. Flying strategic airlifters from out-of-theater MOBs to FOLs could mean that they were arriving with relatively little fuel on board. This could present a significant problem, in that fuel may be in short supply at the FOLs. What fuel is available may be needed to support combat aircraft based at the FOLs. Without sufficient fuel, alternative concepts of operation could be employed. The strategic airlifters could be refueled in flight in the theater, allowing them to touch down at the FOL with enough fuel to return to an MOB. In addition, fuel could be flown in expressly for the strategic airlifters. Both of these alternative concepts of operation would tie up additional assets (tankers or airlifters) that may not be available. In addition, the FOL would likely be in a higher-threat environment that could limit the use of intertheater airlift. Finally, C-5s and C-17s are not able to access airfields as short and soft as those C-130s can use.[7] This could greatly limit the number of usable FOLs. Thus, flying strategic airlifters to FOLs may not be feasible in some theaters and/or scenarios. As a result, this option is not robust and was not carried forward.

Change Theater Routes

Changing the theater airlift routes by using additional FOLs to bring the aircraft closer to the deployed forces could reduce the number of sorties needed to meet a given airlift demand and, subsequently, to reduce the number of C-130s needed to generate them. However, such

---

[7] If payloads are about the same, this short, soft airfield issue may not constrain the larger aircraft. However, to benefit from this option, we must assume that the strategic airlifters are delivering significantly heavier payloads to the FOL per sortie than C-130s.

an improvement would imply that the original routes chosen were at least somewhat inefficient.

While the use of additional FOLs could improve efficiency, the improvement might come at additional risk. Presumably, the choice of FOLs for airlift operations reflected MCS judgment on the threat environment. Changing the airlift routes could therefore incur a different degree of risk and thus may not meet the MCS requirement. Further, while reducing the number of C-130s needed by changing the routes they fly may be possible in the chosen scenarios, it may not be possible in others. As a result, we did not assess changes in theater routes to be a robust solution.

Increase the Maximum Number of Aircraft on the Ground

Increasing the maximum number of aircraft on the ground by employing additional civil engineering units could allow more-efficient use of the C-130s. These units could upgrade the infrastructure at the FOLs, allowing the aircraft to be positioned where they were most needed, rather than where space was available. Improvements in efficiency could subsequently allow fewer aircraft to meet a given airlift requirement. However, the degree to which improvements in the maximum number of aircraft on the ground could improve the operational efficiency of airlift operations is highly theater dependent. Theaters with a large, robust basing infrastructure are unlikely to see much of an improvement, while theaters with very limited basing have much more potential. This solution is thus likely to be highly theater dependent and is, therefore, not very robust.

Increase the Crew Ratio

Currently, the number of hours a C-130 can fly per day is limited by the number of crews available. Historically, this has been 6 hours per day.[8] We therefore also examined increasing the crew ratio. The idea here is that, if the utilization rate of the C-130s is being driven by a lack of crews, the overall number of aircraft for the MCS requirement could be reduced while holding the number of crews constant. As a result,

---

[8]   AMC, 2003, Table 3, p. 12.

the EBH burn rate will remain the same, but recapitalization could be delayed because the fleet could be allowed to fall below 395 C-130s.

The planning factor utilization rate is 6 flying hours per day for sustained operations.[9] Increasing the crew ratio could increase the C-130's utilization rate if no other factors limit its operations. Absent other constraints, a 50-percent increase in the crew ratio could allow the aircraft to operate 50 percent more per day. This, in turn, could allow fewer aircraft to do the work of more, reducing the number of aircraft needed.

In practice, the actual utilization rate is a function of the duration of the flights; the amount of maintenance needed; the refueling, loading, and unloading times; and other factors. For example, a given aircraft could conceivably fly a route with a 2-hour ground time (for loading, fueling, maintenance, etc.) and a 4-hour (total) flight time twice within a 12-hour crew day. The aircraft would thus have a utilization rate of 8 flight hours per day on this route. In contrast, a route with 2 hours of ground time and 2 hours of flight time could be flown three times each day but would only have a utilization rate of 6 flight hours per day within a 12-hour day.

Increasing the crew ratio would help substantially with the first route (a higher crew ratio would allow the aircraft to sustain the higher utilization rate) but would not help at all with the second route (the allowable utilization rate is equal to what the current crew ratio can support, 6 hours per day).

Thus, the potential for changes in the crew ratio to reduce the number of aircraft needed is closely tied to the mix of routes in which the aircraft are used. The mix of routes is, of course, highly scenario dependent. Since reducing the need for C-130s through changes in the crew ratio is robust to changes in the scenarios, we elected to drop this option from further consideration.

---

[9]   AMC, 2003, p. 15. Utilization rate is the average number of flying hours per day for each aircraft type.

Use the Joint Precision Airdrop System

Using JPADS could allow the delivery of materiel and supplies closer to the forward-deployed units and thus could improve the timeliness of the deliveries and reduce the number of trucks needed to ferry supplies from FOLs to these units. The system has worked well in recent operations in Afghanistan and Iraq.

However, much of the cargo (i.e., vehicles) and personnel moved by theater airlift in the demanding transshipment phase of airlift operations may not be suitable for JPADS operations. As a consequence, the potential usefulness of JPADS to reduce the need for aircraft appeared to be limited. In addition, future scenarios might present more-demanding air defense environments, restricting the access of airlifters in forward areas of the battlefield. This would further limit the potential for JPADS to reduce the number of C-130s needed to meet the mission requirement. Given its limited potential and lack of robustness to changes in the scenario, we elected not to carry the JPADS option forward. JPADS is suitable for small cargo deliveries to dispersed operations but is not an efficient way to provide mass delivery to demanding scenarios because of the additional time required for rigging, cost of JPADS units, less-dense loading of aircraft, and probability of loss of cargo due to malfunction.

Increase Army Days of Supply

Army forces typically deploy with three days of supply (DoS). Increasing the number of DoS the Army carried with it could reduce the number of C-130s needed in the sustainment phase of the airlift operation because Army units would not need airlift support as quickly. This, in turn, could require fewer aircraft to meet the mission.

However, this would have a few important implications. Increasing the DoS could increase the number of trucks and trailers needed to carry the supplies. This, in turn, could increase the weight and footprint of the units deployed, thus increasing the amount of airlift needed in the transshipment phase of the operation, which precedes the sustainment phase. Because the transshipment phase requires considerably more airlift than the sustainment phase, increasing the Army DoS would probably increase the number of C-130s needed in the scenario, which is just the opposite of what we were trying to do. We thus dropped this option from further consideration.

Pool Theater Airlift Assets

We also assessed how pooling theater airlift assets might affect the need for C-130s. While the Air Force is the largest operator of C-130s in the U.S. military, it is not the only one. Air Force Special Operations Command (AFSOC) and the U.S. Marine Corps also use the C-130 for airlift. Pooling these resources could improve utilization efficiency and subsequently reduce the need for aircraft.

Unfortunately, these fleets are relatively small, so the potential savings would be very limited. To the extent that these fleets are appropriately sized for their missions, they would not be available for additional airlift missions. Further, taking operational control of AFSOC and Marine Corps assets may not be feasible in some operations because of resistance from within these organizations. This could also encourage these organizations to increase their own demands for additional airlift to make up for the loss of control over their current assets. We dropped this option because of the limited potential gain and negative potential consequences.

CHOP Fewer Aircraft and Improve In-Transit Visibility

The last two options that were dropped in the screening process were a reduction in the number of aircraft CHOPed to other theaters and an improvement in in-transit visibility. Because all the CHOPed aircraft were assigned to a mission and because MCS assumed perfect in-transit visibility, neither of these showed any potential for reducing the number of C-130s needed. As a consequence, we dropped these options.

Observations

We conclude this chapter by noting that two of the options we examined could allow the Air Force to meet the MCS requirement with fewer C-130s. The first was shifting some C-17s to the theater role and using CRAF to fill in for the strategic airlift role. Temporarily shifting C-130s used in training operations to other roles could have a similar effect. However, much of the demand for C-130s is being generated by current operations. The next chapter examines these demands to determine whether the C-130 force could be reduced.

CHAPTER SEVEN

How Current Operations Influence the Demand for C-130s

The previous chapter showed there might be some viable ways to meet the MCS requirement using fewer C-130s. Chapter Three showed that there is a good deal of leverage to delay the need to recapitalize with options of this type. This chapter explores other aspects of C-130 demand—the demand arising from current operations in Iraq and Afghanistan—that are not captured in MCS and explores whether it is possible to realize reductions in C-130 force structure, given the current pace of operations. Even if it is possible to reduce the number of C-130 aircraft needed to meet the MCS requirement, reducing the size of the C-130 fleet may not be desirable. This chapter first evaluates the demand from current operations. Then, as in the previous three chapters, it evaluates potential nonmateriel options for reducing this demand.

In general, the C-130 fleet must meet two distinct sets of requirements: those driven by wartime demands and those driven by peacetime operations. The wartime demand, characterized by the MCS requirement, is a high-intensity, short-term peak that assumes full mobilization of all assets. Virtually all the planned forces are available to meet the challenge identified in MCS. In contrast, during peacetime, much of the force is not mobilized. Peacetime demands are typically driven by a combination of training, channel missions, SAAM, and others. These mission requirements are met by units deploying away from their home stations for relatively short periods (a few days to a few weeks) and are generally met by aircraft stationed within that theater (for example, European channel missions are usually flown by Europe-based aircraft). Aircraft deploy from their home stations, fly the mission, and return to their home stations. There is no need for a rotation base, since these missions are only flown intermittently.

Ongoing contingency operations are different from ordinary peacetime requirements. These missions generally take place in theaters that do not have C-130 units permanently assigned and are typically flown by units deploying to the theater for two to three months at a time.[1] Deploying for a longer period would drive up TDY rates beyond the objective of approximately 120 days per year.[2] If each unit can support operations for two to three months per year, four to six units are needed to support every one that is deployed to these operations.

Figure 7.1 illustrates both these demands. When fewer aircraft are deployed in sustained operations, the MCS requirement drives the size of the needed inventory. However, as the number of aircraft in these contingencies increases much beyond 40 aircraft, the inventory requirement driver is the need to support these operations. The solid lines on the figure indicate the areas where either MCS or sustained operations dominate and drive the force structure requirement.

Aircraft Requirements for Ongoing Operations

For more than five years, the Air Force has been supporting large-scale deployments of C-130s in Southwest Asia. The number of C-130s deployed around the world in contingency operations is presented in Figure 7.2. The initial deployments in support of OEF were followed by a drawdown in the spring and summer of 2002. Toward the end of that year, the deployments increased in anticipation of OIF. The number of aircraft deployed reached a peak of approximately 160 aircraft for

---

[1]   The Air Force has been flying contingency missions in Southwest Asia since 1990. To our knowledge, the Air Force has not sought to base aircraft in the region permanently, largely because the host nations are reluctant to accept a permanent presence.

[2]   Discussions with Headquarters AMC personnel, February 2006. This calculation is discussed in more detail later in the chapter.

Figure 7.1
Sustained Operations Could Drive Need for Forces

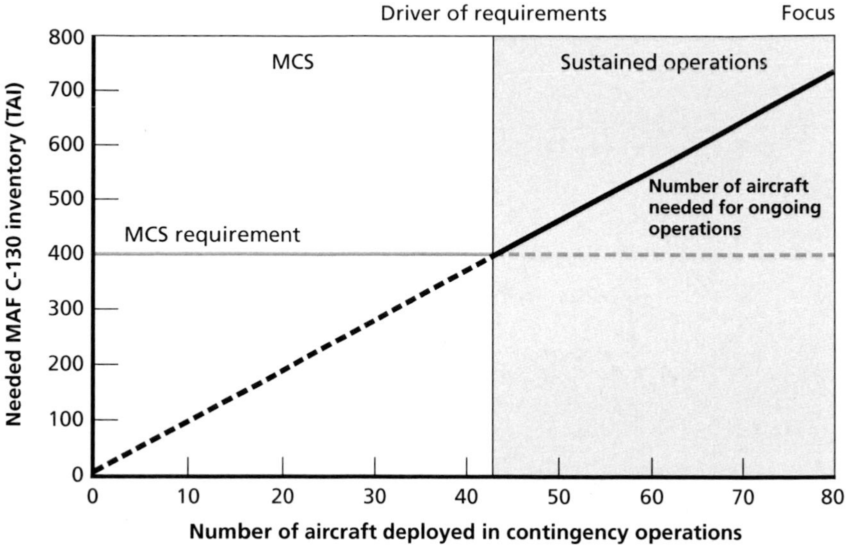

NOTE: Includes all MAF C-130E/H/Js. Assumes a 3:1 dwell-to-deploy ratio and
25-percent ARC participation.

RAND MG818-7.1

much of 2003. In early FY 2004, the number of aircraft deployed fell
off substantially, reaching a plateau of 75 to 76 aircraft. This was maintained through the end of 2005. In 2006, the number of aircraft was
reduced again, falling in two steps, to 46 aircraft. Current plans call
for this force to be maintained indefinitely. In early FY 2007, 40 of the
46 aircraft deployed in contingency operations were in Southwest Asia.

While the OIF peak of 160 was needed for only a few months, the
Air Force has been tasked to support a range of force levels for extended
periods over the last few years. These ranged from a high of about 76
aircraft representing the FY 2004–2005 level to the FY 2007 level of
46 aircraft. In the analysis that follows, we used these levels to represent the demand for ongoing operations.

Figure 7.2
Recent History Was Used to Define the Possible Future Commitment to Contingency Operations

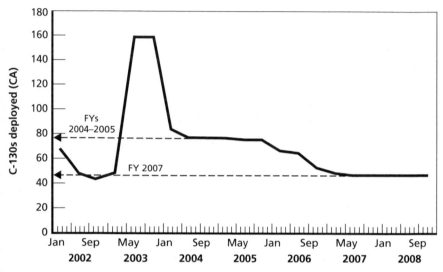

SOURCE: AMC, *Migrating Mobility Forces into AEF*, Scott AFB, Ill., December 2006.
RAND *MG818-7.2*

Force Structure Required to Support Ongoing Operations

The need for forces to support ongoing operations was assessed on a steady-state basis. This has the virtue of providing an accurate picture over an extended period and clearly illustrates the level of activity a given-size force can support. The dynamic picture is considerably more complex, since a given force can support a higher operational tempo for short periods. However, that must be followed by a lower operational tempo to make up for the initial surge.[3]

---

[3] For example, during OIF, several ANG and AFRC units were activated. This greatly increased the operational tempo the mobility forces could support. Over time, the activation authority expired, decreasing the level of operations the mobility forces could support. After a callup authorization has expired, these units cannot be called up for the same operation for two years (U.S. Code Title 10 Subtitle E, Pt. II, Ch. 1209, Sec. 12302).

The assessment was based on the planned FY 2012 active-ARC mix of mobility C-130 forces, which is representative of future plans. The current active force is somewhat larger and thus will be able to support a slightly higher pace of operations for the next few years.

Our steady-state calculations assumed that callup authority has been reserved for future operations and that current operations are being met with the current mix of active forces and with ANG and AFRC participation. In this context, *participation* refers to the share of time that ANG and AFRC personnel are willing to volunteer for duty. We assumed that this level will be about 25 percent, a number consistent with AMC planning factors. However, this level is uncertain because it relies on voluntary participation, rather than a legal obligation, and could change over time, depending on several factors.[4] This uncertainty limits the precision of the estimates of the size of the force needed to support a given level of operations.

We assumed that all mobility C-130 units would be eligible for deployment to contingencies. Thus, no units were required to stay at their home stations to provide lift to theater or CONUS forces. This assumption is consistent with Air Force policy, which has allowed the rotation of virtually all C-130 units into Southwest Asia to support operations in Iraq and Afghanistan.

We also assumed that all units could deploy at their authorized strength. On any given day, nondeployed units are often at less than their authorized personnel and aircraft strength because of leave, training, maintenance, and other activities. As the units prepare to deploy, they are generally able to manage these other demands so that they are able to deploy at authorized strength.

Figure 7.3 illustrates the approach we used to calculate the number of forces that can be supported in contingency operations with a given force size. The number of aircraft available is the total of those provided by the active and reserve components.

---

[4] These could include the demand for pilots in commercial airlines, popular support for the conflict at hand, length of the ongoing conflict (i.e., number of times each is asked to volunteer), and other factors.

Figure 7.3
Estimating the Size of the Force for Meeting the Demands of Ongoing Operations

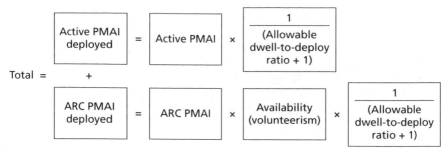

Component	FY 2012 PMAI	Relative Availability	1/ dwell-to-deploy		PAA	PAA/ deployed ratio
Active	114	100%	1/4:1	(0.2)	23	5
ARC	219	25%			11	20
Total	333				34	

(PAA/deployed ratio: ~4×)

The number of aircraft that can be supported in ongoing operations is driven in a large measure by the allowable dwell-to-deploy ratio. This is the ratio of the number of days a given unit spends at its home station (dwell) to the number of days it spends in contingency operations (deploy). Time at home is needed to train aircrews and maintainers, to allow personnel to attend professional military education (PME) courses and attend to other duties, and to provide the desired quality of life for unit personnel. AMC's objective dwell-to-deploy ratio is 4:1. When combined with other requirements, a 4:1 dwell-to-deploy ratio will result in an average TDY rate of about 118 days per year.[5] While units can sustain lower dwell-to-deploy ratios

---

[5]   In addition to time spent overseas flying operational missions, AMC pilots and maintainers go on TDY for upgrade training and PME. They also travel for shorter activities, such as exercises and other training events, conferences, and planning sessions. These can add up to several weeks a year. The actual amount of time spent on TDY for these activities varies by aircraft type. The AMC Aircrew/Aircraft Tasking System, which is used to determine the contribution of crews and aircraft needed from individual units, uses factors of 6 to 13

(and, consequently, higher TDY rates) for significant periods, requiring them to do so has adversely affected training, readiness, and quality of life. With a projected FY 2012 force of 114 aircraft and a dwell-to-deploy ratio of 4:1,[6] the active-duty C-130 mobility forces can sustain approximately 23 aircraft in ongoing operations indefinitely.

The ARC calculation is similar, the only difference being the relative availability of crews. We assumed that, through volunteerism, ARC forces would be available approximately 25 percent of the time.[7] We also assumed that they are treated just like the active units during these periods of availability. With a primary mission aircraft inventory of 219, a relative availability of 25 percent, and a dwell-to-deploy ratio of 4:1, the ARC C-130 forces would be able to provide approximately 11 aircraft to ongoing operations indefinitely.

Using the current force structure and dwell-to-deploy policy, the C-130 fleet can support 34 aircraft in continuous ongoing operations. This is below the lower commitment to ongoing operations of 46 aircraft that was presented earlier in this chapter. The USAF cannot sustain the current level of operations with the current force structure and maintain the desired dwell-to-deploy ratio. Next, we will look at policy options that could allow the Air Force to meet a given requirement for forces in ongoing operations with a smaller force structure.

Potential Options to Increase the Ability to Support Operations

We used DoD's DOTMLPF construct to develop options that could increase the Air Force's ability to support ongoing operations on a sustained basis. These options could affect the size of the active and ARC

---

percent to account for these activities (AMC, *Aircrew/Aircraft Tasking System [AATS]*, Scott AFB, Ill., November 16, 2005, p. 11). The majority of the active AMC aircraft use a factor of 13 percent, which implies a TDY rate of approximately 45 days per year for these other demands.

[6] These 114 aircraft are classified as primary mission aircraft inventory and do not include things like training and backup inventory. The active TAI would be somewhat higher.

[7] This assumption was based on discussions with Headquarters AMC personnel.

forces or their availability to deploy to contingency operations (see Table 7.1).

Change Active-Reserve Mix

The only option we assessed that actually changed the force structure was the shift of some ARC aircraft to AMC. Figure 7.4 shows the size of the force needed as a function of the size of the force deployed in contingencies and the number of aircraft shifted from the ARC to AMC.

Shifting aircraft from the ARC to the active component would increase the availability of those aircraft by a factor of three,[8] reducing the size of the force needed to support a given level of contingency operations. We looked at two cases, one that shifted 24 aircraft and a second that shifted 48 aircraft. These represent approximately 10 and

Table 7.1
Several Options to Increase the Ability of the Air Force to Support Peacetime Contingency Operations Were Considered

Options	Forces	Availability	
		Active	ARC
Doctrine			
Organization	Change active-reserve mix	Remote tours	
Training			
Materiel			
Leadership and education			
Personnel		Reduce dwell-to-deploy ratio	
		Increase number of personnel per squadron	
		Reduce noncontingency TDY	
Facilities			

_____
[8] This calculation assumes an ARC participation rate of 25 percent.

Figure 7.4
Shifting Air Reserve Component Aircraft to Air Mobility Command Could Marginally Reduce the Need for Forces

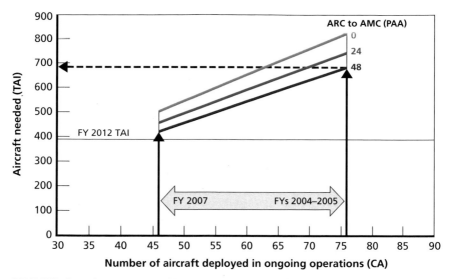

SOURCES: CWB sheet, January 2007; AIRCAT, 2007; *ASIP Master Plan*, 1995; LMAC, 2006; Ramey and Diederich, 2006.
RAND *MG818-7.4*

20 percent of the planned ARC forces. While both options reduce the size of the needed force, neither would allow any actual reductions, since all the force levels needed are in excess of the planned FY 2012 force.

While shifting more aircraft could allow larger reductions in the force structure, such a shift may not be feasible. Many AFRC and ANG C-130 squadrons are the sole tenants at their bases. Shifting these units to active bases could subject the vacated bases to a Base Realignment and Closure (BRAC) action unless other missions are found for them. Neither of these options may be feasible on a large scale, at least in the near term.

Use Remote Tours

The use of remote tours could significantly reduce the burden of supporting contingency operations because one person on a remote tour can support the same level of operations as five people on TDY (with a 4:1 dwell-to-deploy ratio). In this option, a small fraction (5 percent) of the active C-130 aircrews and maintainers would be assigned to a one-year remote tour.[9] These personnel would presumably be midgrade officers and enlisted airmen who had completed most of their qualifications. They would thus not need much in the way of upgrade training, which may be difficult to accomplish during a remote tour. Personnel assigned to these remote tours would spend the year in the region supporting contingency operations. They would replace some crews based in CONUS, Europe, and the Pacific, which would then not have to deploy.

Assigning 5 percent of the C-130 personnel to remote tours in the contingency areas would help by creating a unit that could be assigned to support these contingency operations indefinitely. Being assigned to this unit would mean a permanent change of station, rather than TDY. Personnel on these remote tours would substitute for some of the people deploying from CONUS, Europe, or the Pacific, thus lowering the TDY rates for those units. While deploying to remote tours would involve some personal hardship, the rate we have assumed would impose that hardship very infrequently (i.e., once every 20 years) and thus may be acceptable. This, however, could lead to loss of experienced personnel due to the "seven-day option," which allows eligible airmen to retire or separate rather than take an assignment:

> The "7-day option," which has been in place since 1959, gives enlisted airmen with more than 19 years of service and officers who have passed their initial service commitment the opportunity to turn down a permanent change of station, professional military education or a 365-day temporary-duty assignment within seven days of the tasking, provided they set a firm separation date that falls within a year's time.[10]

---

[9] The FY 2012 programmed force has a total of 114 mobility C-130s in the active component. With a programmed crew ratio of 2.0, there are almost 230 crews. Assigning 12 crews to a remote tour would require just over 5 percent ($12 \div 230 = 5.2$ percent) of the total.

[10] Seamus O'Connor, "'7-Day Option' Gets Second Look," *Air Force Times*, June 6, 2008.

As a result of these difficulties, this option is theoretical and would be difficult to implement.

Remote tours have been commonplace for decades in the combat air forces, and periodic assignment to them has become embedded in the fighter community's "social contract." Personnel go into the career field with the expectation that they be will assigned to remote tours from time to time.[11] Utilizing a similar construct in the air mobility community could significantly increase the availability of the active forces to support contingency operations at a manageable cost.[12]

While this option would not allow reductions in the planned force structure in either case (FY 2007 or FYs 2004–2005), it would lower TDY rates and does not appear to have any significant negative implications. It thus may merit further consideration in force-planning decisions.

This improvement is not without cost. Assigning crews to remote tours imposes a quality-of-life burden, since they will be away from family and friends for 12 months. With 5 percent of the C-130 personnel assigned to remote tours, the average C-130 pilot or maintainer could expect to have one such assignment in a 20-year career. This assignment would be in addition to any other remote assignments in the air mobility community.

Increasing the number of personnel assigned to remote tours beyond the numbers we assumed would provide further benefit but would also impose additional costs. While a detailed assessment of the optimal number of personnel to assign to remote tours was outside the scope of this assessment, the option has potential and merits further consideration.

---

[11] MAF personnel have historically gone on remote tours for staff assignments. Creating a new remote assignment would expand on that considerably.

[12] The relatively large number of models that make up the C-130 fleet makes the use of remote tours somewhat more complicated, in that personnel assigned to remote tours would need to be qualified in the aircraft flown by the units that rotate in and out of the theater. This issue would have to be addressed through scheduling, additional training, or other means. The planned Avionics Modernization Program modification should reduce some of these differences and thus address part of the problem.

Reduce Dwell-to-Deploy Ratio

Reducing the allowable dwell-to-deploy ratio would increase the amount of time personnel could spend supporting contingency operations and thus reduce the size of the force needed to support a given level of operations. For example, shifting from a 4:1 to a 3:1 dwell-to-deploy ratio would reduce the number of forces needed by about 20 percent while increasing the average number of days spent in contingency operations by about 18 per year.[13]

Figure 7.5 illustrates the force sizes needed to support different numbers of aircraft in ongoing operations. Even modest changes in dwell-to-deploy ratio can significantly affect the size of the force needed to support a given deployment level. In this illustration, the active-reserve mix was kept at the same relative percentage as the projected FY 2012 level. Active and ARC units (when volunteering) were

Figure 7.5
Reducing the Allowable Dwell-to-Deploy Ratio Could Reduce the Need for Aircraft

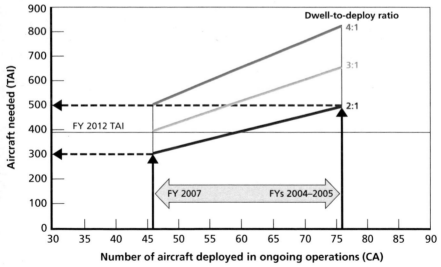

RAND MG818-7.5

---

[13] The numbers presented in this section are an illustration of the option. Air Force personnel with different Air Force specialty codes have different dwell-to-deploy ratios.

assumed to operate at the same dwell-to-deploy ratio. The endpoints for the curves were defined by the number of C-130s in contingency operations—46 in FY 2012 and 76 in FYs 2004–2005.

At a 4:1 dwell-to-deploy ratio, the size of the force needed to support the higher of these two levels is over 800 aircraft. This is considerably larger than the actual FY 2004–2005 force. The Air Force was able to support the higher level of deployments because it was able to extensively utilize ANG and AFRC units that had been called up for OIF. In addition, many active units were operating at much less than a 4:1 ratio.

Reducing the objective dwell-to-deploy ratio to 3:1 would allow the Air Force to support the FY 2004–2005 deployment level with a force of about 650 aircraft. While this is less than the number needed for the 4:1 ratio (more than 800), it is still much more than the planned force structure. Reducing the objective dwell-to-deploy ratio to 2:1 would still leave the Air Force with a requirement for 100 more aircraft than it would plan to have if it were to support the FY 2004–2005 posture indefinitely.

The currently planned number of C-130s cannot even maintain the smaller FY 2007 posture at a 4:1 or 3:1 dwell-to-deploy ratio—although 3:1 is very close. The only case we examined in which the FY 2007 level of contingency operations could be sustained indefinitely and permit a reduction of the planned force was the one in which all the forces operated on a dwell-to-deploy ratio of 2:1. However, operating at a 2:1 dwell-to-deploy ratio would increase TDY rates to almost 170 days per year and could thus affect training, morale, and retention.

The previous illustration assumed that active and ARC units (when volunteering) operated at the same dwell-to-deploy ratio. We also examined what would happen if the ARC operated at higher ratios (see Figure 7.6). We examined three cases. In the first, the active units operated at a 3:1 ratio. That is less than the objective but much higher than the FY 2007 pace, which was about 2:1. We assumed that the ARC forces were able to tolerate a lower dwell-to-deploy ratio, since they would only be involved from time to time. With a 3:1 ratio for the active units and a 2:1 ratio for the ARC, the force needed to support the FY 2004–2005 level would be reduced to about 600. In the second

Figure 7.6
Meeting Demand for Ongoing Operations Requires Significant Changes in Dwell-to-Deploy Ratio

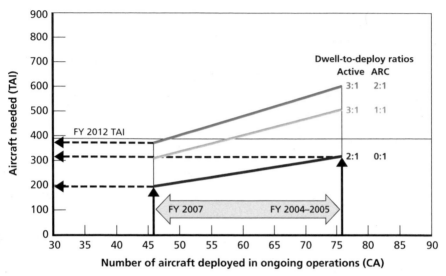

NOTE: Number needed with current active-ARC mix.

case, reducing the allowable dwell-to-deploy ratio for the ARC to 1:1 while keeping the active ratio the same would reduce the force needed to just over 500. All these levels are higher than the planned FY 2012 force structure (see Figure 7.6).

The only case we examined that would allow the Air Force to meet the FY 2004–2005 level of demand while reducing the force structure would be to have the active units operate at a 2:1 ratio, while the ARC operated at 0:1. This may not be feasible. The active units would have to operate with TDY rates of about 170 days per year indefinitely. The implications for the ARC could be much worse: All their volunteer time would be spent in contingency operations. The Air Force depends on reservists choosing to participate, and requiring them to spend all their "volunteered" time overseas could reduce their willingness to participate, making this option infeasible.

If the future level of contingency operations is in line with FY 2007 operations, all three options for changes in dwell-to-deploy ratio would allow reductions in the size of the planned force. However, these reductions would limit the Air Force's ability to sustaining future ongoing contingency operations. This loss would have to be weighed in any assessment of possible reductions.

Increase Number of Personnel per Squadron

Increasing the number of personnel assigned to C-130 units could also increase availability. For a given force size, a larger pool of personnel would be available to generate the same number of personnel deployed in contingencies. This, in turn, could allow a smaller force structure to support the same level of contingency operations. For example, a 50-percent increase in personnel would increase the crew ratio to 3.0. If the needed operational crew ratio remained at 2.0, the Air Force could reduce the size of the force with no loss in capacity to support overseas operations. This could allow the Air Force to reduce the planned force by about 20 aircraft while still meeting the FY 2007 pace of operations. No reductions are possible if the Air Force had to support the FY 2004–2005 pace of operations.

While increasing the number of personnel could allow marginal reductions in the planned force size (at least in one case), it would be a very expensive and inefficient way of doing so. Increasing the number of personnel by 50 percent would increase the number of training hours needed by a similar amount. The annual cost of operating the fleet would subsequently increase by hundreds of millions of dollars. Further, to the extent that the planned crew ratio is well matched to wartime needs, these additional crews would not add much in the way of wartime capability. Finally, increasing the number of annual training hours by 50 percent would greatly increase the EBH burn rate on the current fleet, significantly accelerating the planned retirement schedule for the existing aircraft. This is just the opposite of what the policy option was designed to achieve.

Reduce Noncontingency Temporary Duty

The final option to increase the availability of active forces was to reduce the amount of time needed for noncontingency TDY. This includes activities like off-site training, PME, conferences, and other activities.

Cutting noncontingency TDYs would help by allowing personnel to spend more time in operations for a given TDY level. This could allow them to support a lower dwell-to-deploy ratio and thus reduce the needed force size. However, significantly reducing the number of days available for noncontingency TDYs could greatly limit the training and PME opportunities for C-130 personnel because these activities make up a large share of noncontingency TDY assignments. Limiting the availability of training and PME opportunities of C-130 personnel would very likely stunt their professional growth and, with it, their promotion rates. This, in turn, could limit the appeal of serving in C-130 units and subsequently reduce the quality of personnel they are able to attract.

Because this option appeared to have significant adverse consequences for C-130 personnel and the future quality of their units, we dropped it from further consideration.

Summary of Individual Options

Table 7.2 summarizes our assessments of each of the above options. We examined the potential of the options above to reduce the needed force size under two assumptions about the future need for forces. All the reductions were assumed to be taken from both active and ARC units in proportion to the FY 2012 active-reserve mix.

We assessed each of these policy options according to its potential implications for TDY rates for active forces, potential cost, and other important issues. Options that did not deviate significantly from the outcomes expected under the current plan are shown in green. Those that deviated moderately were colored yellow, and those that were very different were colored red. These evaluations are discussed in more detail later.

The baseline, which illustrates the projected FY 2012 force structure and a 4:1 dwell-to-deploy ratio, is shown in the first row. With a 4:1 dwell-to-deploy ratio, the FY 2012 force does not allow any reductions at either the low (FY 2007) or high (FYs 2004–2005) pace of operations. At a 4:1 dwell-to-deploy ratio, the active units would have an annual TDY rate of just under 120 days. Its costs are those of the

Table 7.2
Summary of Results for Individual Options

| Individual Options | Allowable TAI Reduction[a] | | Potential Implications | | |
	FY 2007 (46)	FYs 2004–2005 (76)	Active TDY	Cost	Other
Baseline (4:1 all)			120		
Change allowable dwell-to-deploy ratio					
All –3:1			138		
All –2:1	90		170		
Active –3:1 ARC 1:1	80		138		ARC[b]
Active –2:1 ARC 0:1	198	77	170		ARC[b]
Other					
Shift 24/48 ARC PAA to AMC			120		BRAC[c]
Increase crew size by 50 percent	22		120		EBH usage[d]
Cut noncontingency TDY by 50 percent			120		Training, PME[e]
At 5 percent of crews on remote tours			101		

Key:
Green, *large scale* (50+)
Yellow, *marginal* (0–50)
Red, *none* (0)

Key:
Green, *positive* or *none*
Yellow, *moderate*
Red, *significant* or *substantial*

[a] Given the pace of ongoing operations.
[b] High ARC utilization might create problems with personnel retention.
[c] BRAC may be needed to allow transfers.
[d] Increasing usage reduces service life.
[e] Reduced training and education opportunities could affect readiness and morale.

planned force. Similarly, since nothing has changed, there are no other implications.

The next four rows show the effects of changes in the allowable dwell-to-deploy ratios. These include shifting both the active and the ARC to 3:1 and 2:1 dwell-to-deploy ratios and shifting the active to a 3:1 and the ARC to a 1:1. Allowing both active and ARC forces to operate at a 3:1 ratio would not allow any reductions at either pace of operations. Operating these forces at a 2:1 ratio would allow a reduction of almost 90 aircraft while still supporting the FY 2007 pace of operations. However, operating the active units at a 2:1 dwell-to-deploy ratio would increase the TDY rate to about 170 days a year. Sustaining this higher TDY rate over an extended period could have important implications for training and morale. This smaller force would not be able to support the FY 2004–2005 pace of operations.

Operating the active units at a ratio of 3:1 and the ARC at 1:1 would allow a reduction of about 80 aircraft. However, no reductions would be possible if the FY 2004–2005 pace of operations had to be maintained indefinitely.[14] Further, deploying the ARC forces at a 1:1 dwell-to-deploy ratio could significantly affect volunteerism and, thus, might not be viable.

The most demanding option we examined was reducing the dwell-to-deploy ratio for the active units to 2:1 and the ARC to 0:1. Doing so could allow large-scale reductions in force structure while still meeting both the FY 2007 and FY 2004–2005 contingency levels. However, implementing this option would have important adverse implications. As we noted earlier, operating the active component at a 2:1 ratio would drive up the active TDY rate to about 170 days a year. Sustaining this pace could adversely affect the force. The 0:1 dwell-to-deploy ratio for ARC under this policy would have its members spending all their volunteered time overseas flying contingency missions. This is in sharp contrast to the current objective, which is to have them spend 20 percent of this time overseas. While it is difficult to estimate precisely how badly this would affect volunteerism, even modest reductions in volunteerism could make this option nonviable.[15]

---

[14] Note that the planned (baseline) force cannot meet the FY 2004–2005 pace of operations either. Supporting that level of operations would require ARC callups for very high dwell-to-deploy ratios for the active units. Reducing the size of the force would make supporting such a deployment even more difficult.

[15] In these calculations, we assumed a volunteerism rate of 25 percent. If that rate declined to 15 percent, the ARC contribution would decline by 40 percent.

The last four rows in Table 7.2 present the remaining options. These all assume that the objective dwell-to-deploy ratio of 4:1 is met. The first of these was to shift 24 to 48 aircraft from the ARC to the active component. As we noted earlier, these shifts would not allow any reduction in the planned force if either of the assumed levels of ongoing operations were to be met. They would have marginal cost effects because active units have somewhat higher operations and support (O&S) costs than ARC units. Because many ARC units are the sole tenants at their home stations, new missions would have to be found for those units if they were to be maintained. A BRAC action would probably be needed if the units were to be deactivated. This could be difficult to achieve on the scale needed to support a significant shift.

Combinations of Options

We also examined combinations of the individual options, or "hybrid" options (see Table 7.3). In the first of these, we looked at the combining a 3:1 dwell-to-deploy ratio with having 5 percent of the C-130 personnel assigned to remote tours. In the second case, we also assumed that 24 ARC aircraft shifted to the active component. The last two cases paralleled the first two, except that the ARC units that had volunteered were assumed to deploy at a 1:1 ratio. While most of the options would allow reductions while still being able to support the lower (FY 2007) range of ongoing operations, none could support the high end of the range. Further, options that included shifting aircraft from ARC to active units on a significant scale could require a BRAC action. Options that included deploying ARC units that had volunteered at a 1:1 ratio may not be feasible if ARC participation falls off in response to the higher deployment tempo.

Options That Reduced Air Reserve Component Forces Exclusively

We examined the same options under the assumption that all the reductions could be taken from ARC (see Table 7.4). Because ARC units are less available to support contingencies than active units, larger reductions in the planned force can be taken while still meeting a given objective. For example, if only the lower (FY 2007) pace of operations has to be met, force reductions of up to 50 percent of the planned

Table 7.3
Potential Attractivness of Hybrid Options

Options	Allowable TAI Reduction[a]		Potential Implications		
	FY 2007 (46)	FYs 2004–2005 (76)	Active TDY	Cost	Other
Baseline (4:1 all)			120		
Hybrid					
−3:1 (all), 5% on remote tours			119		
+ Shift 24 ARC to AMC	22		139		BRAC[b]
−3:1/1:1, 5% on remote tours	61		119		ARC[c]
+ Shift 24 ARC to AMC	87		139		BRAC[b] ARC[c]

Key:
Green, *large* (50+)
Yellow, *marginal* (0–50)
Red, *none* (0)

Key:
Green, *few* or *none*
Yellow, *moderate*
Red, *significant*

[a] Given the pace of ongoing operations.
[b] BRAC may be needed to allow transfers.
[c] High ARC utilization might create problems with personnel retention.

force might be possible. Reductions of this magnitude would require a massive restructuring of the MAF and are neither feasible nor desirable because the remaining forces would not be able to meet the MCS requirement. No reductions are possible if the force was required to meet the FY 2004–2005 pace of operations.

Observations

Operations in Afghanistan and Iraq have tied up large numbers of C-130s over the last few years. While future demands are uncertain, maintaining the capability to provide this level of support greatly limits

Table 7.4
Taking Cuts Exclusively from the Air Reserve Component to Allow Greater Reductions

Options	Allowable TAI Reduction[a]		Potential Implications		
	FY 2007 (46)	FYs 2004–2005 (76)	Active TDY[b]	Cost	Other
Baseline (4:1 all)			118		
Change allowable dwell-to-deploy ratio					
–3:1 (all)			139		
–3:1 Active, 1:1 ARC	97		139		ARC[b]
Other					
Shift 24/48 ARC PAA to AMC			120		BRAC[c]
Remote tours			120		
Hybrid options					
–3:1 (all), remote tours	81		119		BRAC[c]
+ Shift 24 ARC to AMC	153		139		BRAC[c]
–3:1/1:1, remote tours	145		119		BRAC[c]
+ Shift 24 ARC to AMC	169		139		BRAC[c]

Key:
Green, *large* (50+)
Yellow, *marginal* (0–50)
Red, *none* (0)

Key:
Green, *few* or *none*
Yellow, *moderate*
Red, *significant*

[a] Given the pace of ongoing operations.
[b] High ARC utilization might create problems with personnel retention.
[c] BRAC may be needed to allow transfers.

the ability of the Air Force to reduce its force size significantly, given the current makeup of the C-130 forces.

Policy changes could increase force availability and, subsequently, reduce the force size needed to support a given level of operations. However, the only options we found that could allow the Air Force to meet the full range of recent operations have significant adverse implications. These included much higher TDY rates; reduced training opportunities; and a very high pace of operations for ARC forces, which could affect ARC participation and, ultimately, retention. The more-attractive options did not allow the Air Force to support operations at the FY 2004–2005 level on a sustained basis. Some may not even be feasible because they would probably require BRAC actions.

As a consequence, we concluded that the Air Force's ability to postpone the need for new C-130s or SLEPs for existing aircraft is severely constrained by its need to maintain forces to support the demands of ongoing operations. This is unlikely to change in the absence of significant reductions in future demands or in the mix of active and ARC C-130 units.

CHAPTER EIGHT

Cost-Effectiveness Analysis

The most promising policy options that were analyzed in the FSA were subjected to cost-effectiveness analysis. This chapter describes the framework of the cost-effectiveness analysis and the results for each of the policy options analyzed.

Cost-Effectiveness Framework

The base case assumes that the Air Force will fly the MAF C-130 fleet according to current projections.[1] New C-130Js replace older C-130 aircraft as the older aircraft reach the ends of their service lives. We calculated the NPV cost of this case. *Current projections* refers to the annual flying-hour rates and annual EBH accumulation rates projected in the Air Force's Center Wing EBH Report, described in Chapter One.[2]

In each of the policy cases we considered, MAF C-130s have different annual flying-hour and annual EBH accumulation rates than does the base case. In addition, these cases require some other assets and may require purchasing and operating some additional services. We calculated the NPV cost associated with each of the policy cases. The measure of cost-effectiveness of each policy case is the difference between the NPV of the base case and the NPV of the policy case. If this difference is positive, the policy case is a more cost-effective way

---

[1] See Chapter One for the definition of the "MAF C-130 fleet."

[2] More detail on the flying hour and EBH projections will be given later.

to achieve the given level of capability. If the difference is negative, the policy case is a less-cost-effective way to achieve the given level of capability.

We assessed the cost-effectiveness of the following four policy cases:

- *Simulator Case I.* In this case, the Air Force uses simulators to accomplish all basic proficiency training. This reduces the annual flying-hour and EBH accumulation of C-130s from base-case levels by 8 percent and 9 percent, respectively. The utilization rate of C-130 simulators increases, and the Air Force buys one additional simulator.
- *Simulator Case II.* In this case, the Air Force uses simulators to accomplish all basic proficiency training and one-half of all tactical training. This reduces the annual flying-hour and EBH accumulation of C-130s from the base-case levels by 15 percent and 20 percent, respectively. The utilization rate of C-130 simulators increases, and the Air Force buys four additional simulators.
- *Companion Trainer Case.* In this case, the Air Force uses CTAs to accomplish some of the seasoning flying-hour requirement. This reduces the annual flying-hour and EBH accumulation of C-130s from the base-case levels by 13 percent and 10 percent, respectively. The Air Force buys and operates 49 C-12J CTAs in this case.
- *Additional C-17s Case.* In this case, the Air Force uses C-17s to perform those C-130 contingency and channel missions that both (a) have a severity factor greater than two and (b) have a payload greater than 6 tons. This reduces the annual flying-hour and EBH accumulation of C-130s from the base-case levels by 6 percent and 10 percent, respectively. The C-17 fleet flies 5 percent more hours per year than in the base case.

Chapters Four and Six describe these options in substantially more detail. Table 8.1 shows the average annual flying-hour and EBH rates of the MAF C-130 fleet for each case. As discussed later, not all

Table 8.1
C-130 Average Annual Flying Hours per TAI and EBH per TAI in Alternative Cases

Case	Flying Hours per TAI	EBH per TAI	Severity Factor
Base	540	1,150	2.1
Simulator I	490	1,040	2.1
Simulator II	460	920	2.0
CTA	470	1,030	2.2
Additional C-17s	510	1,030	2.0

aircraft fly at the same rate, so these averages simply indicate the overall effects of the policy cases on the flying program of the C-130 fleet.

The Components of Net Present Value

By *NPV*, we mean the value today of all future real expenditures made to provide the given level of capability. Each option we considered will have a different future stream of real expenditures.[3] We discounted future real expenditures by the real discount rate,[4] then converted this future stream into a single value for each option to allow comparison across options.

The NPV cost of military systems, including aircraft and simulators, is the present value of all future life-cycle costs. There are six categories of life-cycle costs:

- research and development (R&D)
- procurement
- military construction (MILCON)

---

[3] By *real*, we mean expressed in dollars of 2006 purchasing power.

[4] We used the real discount rate of 3.0 percent of January 2007, as directed by the Office of Management and Budget for long-term U.S. government investments. (Office of Management and Budget, "Discount Rates for Cost-Effectiveness, Lease Purchase, and Related Analyses," rev., Washington, D.C.: Executive Office of the President, Circular No. A-94, App. C, January 2007).

- O&S
- modification
- disposal.[5]

For our purposes, *acquisition cost* refers to the sum of R&D, procurement, and MILCON, and *sustainment cost* refers to the sum of O&S, modification, and disposal. Some of our cases also include the cost of procuring commercial transportation services. We will detail this later.

The entire NPV is divided into three parts:

- the sustainment cost of aircraft in today's MAF C-130 fleet counted toward the requirement[6]
- the acquisition and sustainment cost of additional C-130 aircraft
- the acquisition and sustainment cost of
  - other kinds of military systems that may be used in the policy option, including simulators, CTAs, or C-17s
  - commercial provision of transportation services.

Evolution of the Mobility Air Forces C-130 Force Structure

According to the Center Wing EBH Report, MAF included 405 C-130E/Hs (TAI) as of January 3, 2007.[7] The report implicitly projects an annual flying-hour rate and an annual EBH accumulation rate for each of these 405 C-130E/Hs. The report also explicitly supplies the cumulative total EBH each had accumulated as of January 3, 2007, the projected date by which will have accumulated 45,000 EBH, and the number of flying hours it will have accumulated by that

---

[5] Disposal costs are a very small part of total costs, but the Air Force guidance for this kind of study specifies including them for completeness. See Air Force Materiel Command, *Analysis Handbook, A Guide for Performing Analysis Studies: For Analysis of Alternatives or Function Solution Analyses*, Kirtland AFB, N.M.: Office of Aerospace Studies, July 2004.

[6] In this case, those needed to meet the MCS requirement.

[7] As noted previously, the Center Wing EBH Report is a specific spreadsheet drawn from the C-130 System Program Office's AIRCAT database on January 3, 2007.

date. Annual flying-hour and EBH rates follow from those. The NPV analysis includes the projected cost of operating and sustaining those among these aircraft that counted toward the requirement.

In addition, the Air Force had 37 MAF C-130Js on January 3, 2007. The most recent budget documents project that the Air Force will acquire an additional 28 MAF C-130Js by the end of FY 2010.[8] Of course, actual procurements are likely to turn out somewhat differently as future sessions of Congress rework budgets. We took the projection of 28 as a given.

The size of the MAF C-130 fleet at the time of this analysis is 442 (405 C-130E/Hs plus 37 C-130Js), which exceeds 395. For the NPV analysis, we defined "C-130E/H models counted toward the 395-requirement" as those that are needed, given the number of MAF C-130Js in the inventory, to reach a TAI of 395. We included the costs associated only with these C-130E/H models in the NPV totals. Therefore, costs associated with any other C-130E/H aircraft (i.e., in excess of the quantity "395 minus the number of MAF C-130Js") are excluded from the NPV calculations. Table 8.2 shows how many C-130E/H models are counted toward the 395-requirement in each year until all are retired.

In the base case, all the 405 C-130E/H models fly at their unique annual flying-hour and EBH rates, as projected in the Center Wing EBH Report. In the analysis, they can contribute to the requirement until they reach 45,000 EBH, at which point they are retired. (As noted earlier, this assumes that all of the 405 pass TCTO 1908 so that they can continue to fly unrestricted beyond 38,000 EBH.) When, as a result of these retirements, the number of C-130E/H aircraft falls to 330, the total C-130 fleet, equal to these 330 plus the 65 existing or projected additional C-130Js, just equals the requirement. As additional C-130E/Hs are retired beyond that, new C-130Js are acquired to keep the total fleet size at 395.[9] Table 8.2 projects the future of the MAF C-130 fleet in the base case through FY 2046.

---

[8] The projection is based on U.S. Air Force, 2007b.

[9] This analysis does not include aircraft attrition.

Table 8.2
Evolution of the MAF C-130 Fleet in the Base Case

End of FY	E/H Series		J Series
	Total	Counted Toward Requirement	
2007	404	351	44
2008	397	343	52
2009	391	336	59
2010	378	330	65
2011	364	330	65
2012	347	330	65
2013	329	329	66
2014	313	313	82
2015	298	298	97
2016	282	282	113
2017	274	274	121
2018	267	267	128
2019	250	250	145
2020	242	242	153
2021	236	236	159
2022	230	230	165
2023	223	223	172
2024	214	214	181
2025	209	209	186
2026	203	203	192
2027	195	195	200
2028	190	190	205
2029	182	182	213
2030	173	173	222
2031	155	155	240
2032	141	141	254
2033	121	121	274
2034	117	117	278
2035	108	108	287
2036	98	98	297

Table 8.2—Continued

| End of FY | E/H Series | | J Series |
	Total	Counted Toward Requirement	
2037	84	84	311
2038	71	71	324
2039	54	54	341
2040	39	39	356
2041	31	31	364
2042	26	26	369
2043	23	23	372
2044	7	7	388
2045	1	1	394
2046	0	0	395

Highlights of the fleet evolution in this analysis are

- The first C-130J beyond the 65 either existing or currently planned is delivered in FY 2013. This is also the year (necessarily, by the way that we have structured the analysis) in which all MAF C-130E/H aircraft are counted toward the requirement.
- Annual C-130J deliveries average about ten between FYs 2013 and 2046. In the analysis, deliveries occur precisely when C-130E/H aircraft hit the 45,000-EBH limit and are therefore necessarily choppy. In reality, one would expect some smoothing. However, for this analysis, we elected not to exogenously constrain delivery patterns because such constraints are necessarily judgmental and could affect the results arbitrarily.
- The last MAF C-130E/H is retired in 2046.

C-130J aircraft that are procured when the total MAF C-130 falls below the 395-requirement each fly at the annual flying-hour and EBH rates of the aircraft they replace. The 65 C-130Js that either exist now or we project will be procured by the end of FY 2010 fly at the rates of the 75 C-130E/H aircraft that will be retired before the total MAF C-130 fleet hits the 395-requirement. This analytical approach keeps the overall fleet future flying-hour and EBH rates consistent with today's projections as the fleet evolves from C-130E/Hs to C-130Js.

In the alternative cases, the annual flying-hour and EBH rates of the MAF C-130s are lower, as shown in Table 8.1. This means that the existing C-130E/H aircraft reach 45,000 EBH later and, thus, that the procurement of C-130Js (beyond the 28 projected by FY 2010, based on budget documents) is postponed. In earlier chapters, we discussed how this would affect the quantities and mix of the fleet at any given time.

Elements of Cost Estimates

This section presents our estimates of all the cost parameters used in the NPV calculations, beginning with acquisition costs, then sustainment costs. As an overall summary, Table 8.3 shows, for each military system or other item included in the NPV, the estimated flyaway cost and the O&S cost in the first year of operation. These are the primary components of acquisition and sustainment costs, respectively.

Acquisition Cost

We begin with the flyaway cost of each of the systems, which is the largest component of acquisition cost.

Table 8.3
Flyaway and Operating and Support Costs

Military System or Other Item	Flyaway Procurement Cost ($M)	O&S Cost per Hour ($000)
C-130E	—	14.8
C-130H	—	14.2
C-130J	61	11.0
C-17	197	19.9
CTA (C-12J)	4	3.6
Simulator	23	0.7
TWCF hour	—	5.8

NOTE: Costs are in 2007 dollars.

C-130J. The FY 2007 President's Budget, submitted in winter 2006, shows procurement of 11, 8, and 9 C-130J aircraft in fiscal years 2005, 2006, and 2007, respectively.[10] The average unit flyaway cost for those is between $67 million and 70 million.

In December 2006, the FY 2006 GWOT supplemental authorization approved a $256.2 million (then-year) contract for four C-130Js for delivery in 2010. The delivery date implies a procurement date of FY 2008 or FY 2009. The deflated FY 2007 unit cost is $61 million. Since this represents the most recent negotiated price for the C-130J at the time of our analysis, we used this latest figure of $61 million as our estimate of the unit flyaway cost for future buys.

In addition, our discussions with LMAC were consistent with an expectation that this most recent price would continue into the future, at least as long as the Air Force continued to purchase in the neighborhood of ten MAF C-130Js annually. As the discussion of Table 8.2 indicated, this is about the rate we expect. We would also expect some decrease in the price if the annual buy were to increase substantially. Discussions with Lockheed representatives indicate that the plant is currently tooled for 24 per year, and they state that unit costs would fall at production rates higher than those implied in the U.S. budget we used as a baseline.

C-17. Our source for C-17 procurement cost was the C-17 Selected Acquisition Report (SAR) dated December 2005. The $197 million flyaway cost is the SAR figure for FY 2007, with a buy of 15 C-17s. In that SAR, the last buy is in FY 2007, at about a 25-percent higher unit cost. We use the lower FY 2006 cost because the last year of production has historically shown higher costs as production winds down.

Companion Trainer. We use the C-12J as our analogue for estimating the cost of a CTA. The source of the estimate is AFI 65-503 Table A-10 (Unit Recurring Flyaway Cost). It shows a unit recurring flyaway cost of $4.2 million for the C-12.

---

[10] Air Force Financial Management and Comptroller, U.S. Air Force, *Committee Staff Procurement Backup Book: FY 2007 Budget Estimates*, Vol. I, *Aircraft Procurement*, Washington, D.C.: U.S. Air Force, February 2006, p. 2-19.

Simulator. The FY 2007 President's Budget for simulator procurement costs shows a recent procurement cost for a weapon system trainer of $23 million. Personnel in the Training Division at Headquarters AMC confirmed that this is also their expectation of future prices. We note that this has dropped since earlier buys; the initial purchase cost had included substantial nonrecurring costs. Because this represents that most recent budget-quality simulator cost estimate, we use it as our estimate of the unit cost for future buys. Personnel in the Training Division at Headquarters AMC also confirmed that this is consistent with their expectation of future prices.

Other Parts of Acquisition Cost. Besides the flyaway cost, the average unit procurement cost (AUPC) of an aircraft includes several other cost categories, including systems engineering and program management, training, data, initial spares and repair parts, and peculiar support equipment.

We estimated the total additional cost associated with these categories as a proportion of flyaway cost. The estimated factor is 16 percent, based on information in FY 2001–2007 Air Force budgets and in the 2005 SARs. We use this factor to escalate flyaway cost to AUPC for the C-130J, C-17, CTA, and simulators.

Because these systems already exist, we did not include R&D costs. We did include a MILCON cost for buildings to house new simulators beyond those the Air Force already owns. AMC's Director of Operations (AMC/A3) informed us that the recent construction cost for a building to house one simulator and associated rooms at Little Rock was $4.7 million. We applied this to all new simulators.

Sustainment Cost

We begin with the O&S cost of each of the systems, which is the largest component of sustainment cost. We first discuss our approach for modeling aircraft O&S costs, which is primarily based on the Cost-Oriented Resource Estimating (CORE) model. We then address the O&S costs associated with simulators and our treatment of TWCF costs, as well as our estimates of the other parts of sustainment cost, which are modifications and disposal.

Aircraft O&S Projections Based on the Cost-Oriented Resource Estimating O&S Model. O&S costs for all the aircraft were estimated using a tailored version of the CORE model available through the Office of the Secretary Air Force, Financial Management and Comptroller.[11] The CORE model estimates yearly squadron direct costs and the aircraft's share of indirect costs in the standard DoD cost element structure for O&S costs. The CORE model is intended to capture and estimate the aircraft's total ownership cost during the O&S phase. From these yearly costs, costs per aircraft or per flying hour can be calculated. The seven main O&S cost elements are

1. Mission Personnel (including crew and maintenance)
2. Unit-Level Consumption (including fuel, consumables, and repair parts)
3. Intermediate Maintenance
4. Depot-Level Maintenance (including airframe and engine)
5. Contractor Logistics Support
6. Sustaining Support
7. Indirect Support (including initial skill training for replacement personnel and the weapon system's share of base operating support).

The model requires inputs specific to an aircraft MDS for factors that drive major elements of O&S costs. For example, the model uses the number of authorized maintenance personnel per squadron and their average yearly compensation to estimate mission maintenance personnel costs. It uses fuel consumed per flying hour and fuel cost per gallon to estimate fuel consumption costs.

The model treats some costs as being fixed each year, while some costs vary with usage. For example, mission personnel are fixed. Fuel consumption, consumables, depot-level reparable, and engine overhaul costs vary directly with flying hours. This fixed and variable treatment could be modified, which would be advisable especially if the estimated number of flying hours per aircraft per year differed a great deal from

---

[11] The CORE model is generally used in the Air Force for analyses of aircraft O&S costs of this nature.

historical experience. In this analysis, we assumed flying-hour or usage rates similar to those of recent experience and so did not change the treatment of fixed and variable costs in the CORE model.

The Office of the Secretary Air Force, Financial Management and Comptroller, publishes CORE model inputs as attachments to AFI 65-503. Many of the CORE model inputs are based on recent actual costs, and using them in the model usually results in estimates that are close to actual costs when the same numbers of aircraft and flying hours are compared.[12] However, CORE estimates can differ markedly from actual costs in two particular elements—mission personnel and indirect costs. In the following paragraphs, we discuss the reasons for these differences.

An important CORE model input for mission personnel costs is the authorized number of people per squadron. In actual practice, the Air Force assigns fewer or more personnel to squadrons than authorized for a variety of reasons. The number of assigned personnel determines actual costs, which are reflected in AFTOC. So there are often differences between the costs in AFTOC and the results of the CORE model for the same scenario because of differences in the number of authorized and assigned personnel.

With one exception, we modeled mission personnel costs using CORE inputs and, thus, authorized manning levels. These represent the professional assessments of Air Force personnel analysts and are Air Force policy on how many personnel are required to carry out the mission effectively. These levels are thus appropriate indicators of the kinds of long-run costs that are relevant for our analysis. The particular circumstances of any year's personnel assignment outcomes should not, in our judgment, be used for these projections. Each year's outcome reflects the relatively slow adjustment by the Air Force to the authorization target, based on specific accessions, deployments, reorganizations, and so on, while the target itself is the right measure of the long-run costs.

---

[12] When we say "actual costs" here, we mean precisely the level of costs reported in the Air Force Total Ownership Cost (AFTOC) database.

The one exception is the C-130J, for which no formal analysis of authorized maintenance personnel has yet been done by the Air Force, and for which CORE uses, by default, authorization levels for C-130E/ Hs. The formal analysis to determine C-130J maintenance manning levels will be done at AMC using the Logistics Composite Model. In FY 2006, the C-130J had significantly better maintenance-manpower-per-flying-hour and mean-time-between-maintenance metrics than the C-130E/H, based on information we received from the C-130J USAF Sustainment Program Manager and from AMC logistics personnel. In addition, based on AFTOC, it had significantly lower maintenance manpower costs than the C-130E/H in ANG and ARC squadrons. As we discuss later, we assess that the 2006 experience is a valid basis for projection of future long-run costs, and therefore we judge that this is a better basis for our analysis than the C-130E/H authorized manpower levels. Thus, the manning levels we use for the C-130J are based on the 2006 AFTOC data, not on the CORE AFI 65-503 inputs.

The second element in which there are often differences between the result of the CORE model and AFTOC is indirect support. The indirect support element contains subelements for personnel support, which includes specialty training, permanent change of station, and medical support costs; and installation support, which includes base operating support and real property maintenance costs. The CORE model often produces higher costs for specialty training of replacement personnel than is captured in AFTOC. For installation support, the costs in AFTOC can reflect the peculiarities of basing modes because the installation costs of a base are apportioned to the aircraft at that base. So, for example, if a C-130E squadron is the sole tenant at a base, all the base's installation costs will be apportioned to those few aircraft, whereas another C-130E squadron that is one of several tenants at a base will have a much smaller share of installation costs. Since a substantial portion of installation support costs is fixed, these costs tend not to grow in proportion to the number of aircraft on a base. Thus, the AFTOC-reported installation cost per aircraft will depend on the specifics of the basing structure. We used the CORE inputs rather than the AFTOC actual values for these indirect costs because they represent the results of Air Force analysis of how these costs vary across the entire Air Force as force structure changes. The AFTOC entries can represent accounting conventions that are peculiar for each weapon system and may not reflect all the factors that the CORE model inputs properly account for.

For the cost elements unit-level consumption, depot maintenance, and sustaining support costs, with one exception we used the actual costs in FY 2006 as reported in AFTOC rather than the inputs found in AFI 65-503. The exception is that for PDM for the C-130E and C-130H, we used actual cost figures provided to us by the depot at Warner Robins AFB. We judge that these have better fidelity in distinguishing differences in true costs between the two models.

There are two reasons for using the actual costs from AFTOC, or directly from the depot, rather than AFI 65-503 inputs for the cost elements unit-level consumption, depot maintenance, and sustaining support costs. First, and most importantly, two of the aircraft—the C-17 and C-130J—have just transitioned from interim contractor support funded by the aircraft procurement appropriation to contractor logistics support (CLS) funded by operating accounts, so these cost elements are included in AFTOC in FY 2006. Inputs in AFI 65-503, based on costs from prior years, would not represent these costs as well as the FY 2006 AFTOC data do.

Second, the Air Force has a small but slowly growing fleet of C-130Js. In the first few years that a weapon system is fielded, it is common to experience inefficiencies and associated distorted costs. This results from a small fleet bearing significant fixed cost that can support a much larger fleet, as well as learning curve issues. Based on discussions with knowledgeable Air Force and industry persons, it is our assessment that FY 2006 is the first year in which the actual data make a good basis for projecting the long-run future O&S cost of the C-130J. In addition, we compared AFTOC-reported costs for C-130H and C-130J unit-level consumption (fuel, consumables, and depot-level repairables) with Coast Guard data on the same costs for its C-130Hs and C-130Js. The Coast Guard costs per flying hour in these categories for each aircraft series were close to our calculation for the Air Force aircraft. This is consistent with our assessment that projections based on AFTOC FY 2006 costs are appropriate. The Coast Guard data are based on the U.S. Coast Guard Operations and Maintenance Plan and discussions with the C-130 System Manager in the U.S. Coast Guard.

There are two additional cost elements besides the five just discussed, which are intermediate maintenance and contractor support. None of the aircraft in this analysis incur any intermediate maintenance costs. AFTOC reports the value of CLS in a program as a single value in the element for CLS costs. There is no insight into the nature of the CLS costs in AFTOC. The standard CORE methodology mirrors the reporting in AFTOC, so that the analyst enters the CLS cost per aircraft or per flying hour.

Three of the aircraft in this analysis—the C-130J, C-17, and C-12—use CLS. The CLS in these three programs includes maintenance personnel, spare and repair parts, airframe and engine overhaul, and sustaining support. We model the CLS costs in the elements where the support is actually provided because some elements are variable with usage and some elements are susceptible to cost growth over time. We apportioned the total CLS costs among the appropriate elements based on the CLS brochures for these programs.[13] For example, for the C-130J, we used the FY 2008 CLS brochure, which was prepared in the middle of FY 2006 and contains budgeted costs and flying hours for that year. Although it does not represent actual costs, it provides a good representation of the contractual arrangement for C-130J CLS. Much of the cost is for propulsion system "power by the hour" and airframe spare and repair parts unique to the C-130J. We verified actual obligations for FY 2006 with the C-130J USAF Sustainment Program Manager. Similar procedures were used for the other CLS aircraft. For all aircraft, these allocated CLS costs are additive to the other costs in the specific cost elements that were discussed previously.

For all aircraft, age-related cost growth was included for some sustainment cost categories. Linear cost growth rates were 1 percent for consumables, 2.4 percent for depot-level repairables, 2 percent for

---

[13] A CLS brochure is a budget request prepared by system sustainment managers for weapon systems that use CLS. The brochure provides more-detailed information on CLS costs and tasks than is available in the Air Force's O&S cost database. (Department of the Air Force, Contractor Logistics Support Brochures, various programs, prepared for FYs 2006, 2007, and 2008.)

PDM of C-130E and C-130H, and 1 percent for PDM of C-130J, C-17A, and C-27J. These estimates are based on the statistical analysis of historical O&S costs of cargo and tanker aircraft done by RAND Project AIR FORCE for the KC-135 Recapitalization AoA.[14] As discussed more fully in that report, the lower aircraft PDM growth rate for newer aircraft reflects improvements in aircraft design for supportability and the resulting expectation that these aircraft will age more gracefully.

One final issue is that, for the C-130E and the C-130H, AFTOC includes, for the first time in FY 2006, direct costs for GWOT TWCF hire of other aircraft as a direct charge against the C-130s. This of course is not properly part of C-130 O&S cost for this analysis and, in concurrence with AMC financial management personnel, we do not include these costs in any of our sustainment cost estimates for the aircraft systems.

Simulator. The ongoing O&S of the trainers, including developing and maintaining courseware, maintaining the training devices, and providing instruction, is provided through CLS. We calculated the hourly O&S cost of the trainers from information in the C-130J Training System CLS Brochure. New simulators require buildings in which to house them. We estimated annual building maintenance costs using the appropriate cost factor for a training facility in the DoD Facilities Pricing Guide.[15]

TWCF Hours. In the CTA alternative, one of the costs is for procuring the TWCF hours not flown on C-130s elsewhere. We estimated their cost as $5,750 per hour, based on AFI 65-503.[16] This is the average of the C-130E, C-130H, and C-130J rates applicable to DoD. These rates are meant to approximate the cost of alternative sources of transport so that other DoD users have the appropriate incentives in their

---

[14] Reported in Michael Boito, Gregory G. Hildebrant, Fred Timson, *Analysis of Alternatives (AOS) for KC-135 Recapitalization: Appendix F—Operating and Support Costs*, Santa Monica, Calif.: RAND Corporation 2005, Not Available to the General Public.

[15] DoD, "Unified Facilities Criteria (UFC): DoD Facilities Pricing Guide," Vers. 7, March 7, 2007. This guide has since been superceded.

[16] AFI 65-503, 1994.

choices between organic and contract transport services. Given this, we judge them to be the best indicator of the cost of obtaining the hours in the private market.

Other Parts of Sustainment Cost. In our analysis, all aircraft systems incur a base constant level of modifications, in real dollars, per year. This represents a "level-of-effort" analysis modeling approach. We cannot forecast well what modifications will occur more than a few years into the future, but we included this level of effort to capture the fact that different systems tend to incur different average annual levels of cost. For the C-130E, the C-130H, and the C-17, we projected forward the average of the last five years' actual modification costs per aircraft from the AFTOC Appropriations data file to approximate annual costs of $86,000; $125,000; and $1,170,000, respectively (in 2007 dollars). We projected the C-130H figure for the C-130J, then estimated the CTA figure using the same fraction of its procurement cost as for the C-130J, yielding a CTA estimate of $9,000 per year. This level-of-effort estimate is assumed to capture future SLEP costs as well. Another cost-estimating option, recommended by one of our reviewers, would be to include one or more discrete SLEPs at given years for specific aircraft. We elected not to do this because of the great uncertainty about the nature of these future SLEPs, making such SLEP cost estimates essentially arbitrary. Moreover, such estimates would increase all aircraft and other system costs, so not estimating specific SLEPs helps prevent bias in comparisons across aircraft programs.

We did include modification costs for the ongoing Avionics Modernization Program for the C-130E/H. This program has suffered significant cost growth recently, and projections are necessarily very uncertain, given the recent instability of the actual costs. From discussions with Air Force cost analysts, we project that the average unit cost will be $13 million in 2007 dollars, about twice that of the most recent SAR. Given this high recent cost growth, we project that only C-130E/Hs that remain in the inventory beyond 2020 will receive this modification. Again, this program's future is currently quite uncertain.

No modification costs were included for simulators, since these costs are included in the O&S cost estimate. The O&S estimate primarily includes payments for CLS, which includes upgrades. Disposal costs are estimated as two 1,000ths of production cost.[17]

Cost-Effectiveness Analysis Results

For the base case, we calculated the NPV of all future expenditures that must be made to provide the 395-aircraft capability level of intratheater transport. This NPV includes the sustainment costs for the 395 existing C-130s that count toward the requirement and the acquisition and sustainment costs for the new C-130Js to be acquired to replace older aircraft as they retire. To account for differing service lives, we also included a sinking fund for each new purchase to provide for its replacement. That is, if two assets have the same purchase price and annual sustainment cost but different service lives, the one with the shorter service life will be assessed a larger annual sinking fund payment to account for the fact that its replacement will occur earlier.

Another way to view the NPV of all future expenditures is that it is the sum of money that, if invested now at 3-percent real interest, would yield revenues sufficient to fund all the acquisition and sustainment costs of the entire future fleet. The base-case NPV is about $100 billion. Of this, about half is the present value of the cost of operating the existing C-130 fleet until retired, and the other half is the present value of all future costs of acquiring and operating additional C-130Js.

The Simulator Cases Are Cost-Effective

Simulator Case II utilizes an additional 35,000 hours of simulator time over the base case, requiring four additional simulators. At the O&S cost of $725 per hour, the additional 35,000 hours cost about $25 million per year, which, at 3 percent, is equivalent to an NPV of about $850 million. The AUPC cost for a simulator is $33 million; this analysis assumes procurement of a replacement simulator every 20

---

[17] General Accounting Office [now the Government Accountability Office], *Financial Management: DoD's Liability for Aircraft Disposal Can Be Estimated*, Washington, D.C., GAO/ AIMD-98-9, November 1997.

years (conservative). The present value of the sinking fund needed to pay for future replacements is $41 million. Thus, the present value of the gross costs associated with this case is about $1.1 billion. However, the NPV of the C-130 fleet needed to meet the requirement decreases by about $7.8 billion, so there is about a $6.7 billion benefit in this case.

The largest factor in the $6.7 billion is that each aircraft flies about 80 fewer hours per year. The marginal cost per flight hour is about $5,000 per hour across all C-130 models,[18] so this leads to annual savings of about $400,000 per year per aircraft, or about $160 million per year for the fleet of 395. The present value of this is about $5.3 billion. The remaining $2.4 billion is the present value of postponing the replacement of the aircraft because they fly more calendar time before hitting the 45,000-EBH limit.

Simulator Case I utilizes an additional 20,000 hours of simulator time over the base case and requires one additional simulator. At the O&S cost of $725 per hour, the additional 20,000 hours cost about $14 million per year, which, at 3 percent, is equivalent to an NPV of about $450 million. The AUPC cost of a simulator is $33 million; this analysis assumes procurement of a replacement simulator every 20 years (conservative). The present value of the sinking fund needed to pay for future replacements is $41 million. Thus, the present value of the gross costs associated with this case is about $0.5 billion. However, the NPV of the C-130 fleet needed to meet the requirement decreases by about $4 billion, so there is about a $3.5 billion benefit in this case.

The largest factor in the $4 billion is that each aircraft flies about 45 fewer hours per year. The marginal cost per flight hour is about $5,000 per hour across all C-130 models, so this leads to annual savings of about $225,000 per year per aircraft, or about $90 million per year for the fleet of 395. The present value of this is around $3 billion. The remaining $1 billion is the present value of postponing the replace-

---

[18] The marginal cost is less than the average cost shown in Table 7.3 because not all parts of cost vary with flying hours in our analysis. For example, in our treatment, squadron operations and maintenance manpower is determined by wartime tasking levels, so it does not vary with flying hours.

ment of the aircraft because they fly more calendar time before hitting the 45,000-EBH limit.

The Companion Trainer Case Is Not Cost-Effective

In this case, 49 C-12J aircraft are procured. The requirement is 42 PAA; we added seven to reflect an availability factor of 85 percent. In our analysis, these aircraft fly 1,200 hours per year per PAA at a cost of $3,600 per hour, or $4.3 million per year. The present value of this for the fleet of 42 PAA is about $6 billion. The AUPC is $5 million, and the present value of the sinking fund is an additional $4 million. Thus, the present value of the acquisition cost for the fleet is about $450 million, and the gross cost of the aircraft acquisition in this case is about $6.5 billion. In addition, because the hours C-130s do not fly in this case are assumed to be TWCF hours, the cost of procuring them elsewhere must be included. At $6,000 per hour, with about 30,000 flying hours lost, the present value cost of this is about $5 billion. So the total gross cost of this case is between $12 billion and 13 billion.

The gross savings for the C-130 fleet in this case are only around $5.5 billion, so this policy incurs a total NPV loss of about $7 billion. In terms of flying hours per year and EBH per year saved per C-130, this case is about halfway between Simulator Cases I and II, so its gross savings are also about half way between them.

The C-17 Case Is Not Cost-Effective

In this case, 32 additional annual flying hours per C-130 must be flown by C-17s, or a total of about 12,500 C-130 hours per year. However, C-17s will not need that many flying hours to perform the equivalent transportation service. The C-17 has a 28-percent block speed advantage at the average sortie distance of a C-130 (500 miles), so no more than 9,000 hours would be needed.

The cost of adding one hour per year to a C-17's flying schedule is $18,000. This includes the marginal flying-hour cost of about $12,000 per hour plus a capital charge of about $6,000. The capital charge is based on a 35,000-hour flying-hour life for the C-17 and on the present value cost difference between retirement schedules based on 1,001 versus 1,000 hours per year. As a roughly indicative intuitive measure, dividing the AUPC of $235 million by the 35,000-hour life approximates the capital charge reasonably well.

The 9,000 additional hours per year at $18,000 per hour lead to an additional cost of $162 million per year. The present value of this is about $5.5 billion, which is the gross cost of this policy option.

The gross savings of the C-130 fleet in this case are only around $3.5 billion, so this policy incurs a total NPV loss of about $2 billion. The average of the percentage reductions in flying hours per year and EBH per year per C-130 are about the same as in Simulator Case I, so the gross savings are also about the same.

Conclusions

We estimated the cost consequences, in NPV, of four alternative policies that would extend the lives of the C-130 fleet. We found that increasing simulator use had NPV benefits, about $7 billion in the maximum simulator use case. This is equivalent to a net saving of $210 million per year. The actual pattern of outlays and gross savings would not be constant over time but would fluctuate as annual purchases change. However, this is the relevant average annual saving for cost-effectiveness considerations.

We found that neither a CTA program nor moving high-EBH missions to the C-17 fleet would be cost-effective. In these cases, the additional costs would outweigh the benefits of fewer flying hours and longer life of the C-17s.

CHAPTER NINE

Conclusions

The C-130 fleet performs critical air mobility functions for the nation, but part of that fleet is at risk because of age-related factors. The oldest aircraft, C-130Es and H1s, are at greatest risk from structural fatigue damage, corrosion, and aging aircraft systems. Uncertainties about the structural health of individual aircraft further complicate management of the aging fleet. CWB fatigue cracking is the most immediate life-limiting issue, but other aging issues are looming. If the Air Force chooses to retain older C-130s, it will also need to address functional system issues.

The risks of catastrophic structural failure increase markedly when flying aircraft beyond 45,000 EBH without structural mitigation. Moreover, there is also considerable uncertainty about the structural health of individual high-EBH aircraft because of limitations in charting the historical accumulation of fatigue damage and in the ability of inspections to reliably discover fatigue damage on aging aircraft.

CWB structural fatigue will cause the number of available C-130s to fall below the MCS requirement by 2013. This assumes that the aircraft accumulate EBH at the same level as recent historical experience. During this study, we looked at many potential nonmateriel solutions to close the capability gap. Our analysis identified three broad classes of solution options.

The first class of options reduces the rate of EBH accumulation. Although we identified several options that could significantly reduce EBH accumulation on the aircraft, none significantly delayed the need to recapitalize the C-130 fleet. Much of the C-130 fleet is composed of high-EBH aircraft having only a few years of life remaining at the current operational tempo. Even a fairly significant reduction in EBH accumulation will delay the need to recapitalize only by one or two years.[1] Most options we analyzed that could provide a year or two of delay cost more on an NPV basis than buying new aircraft. The only potential option in this class that resulted in an NPV savings was increased use of simulators. Although this option can delay recapitalization only by a year or two, this option has a significant NPV savings of about $7 billion—that is, about $200 million per year.

The second class of solution options increases the amount of EBH available. This would mean flying the aircraft beyond the 45,000-EBH limit. There are essentially two ways to do this. The first is a SLEP to mitigate accumulated fatigue damage on critical structures, essentially rolling back accumulated EBH; the second is flying the aircraft beyond 45,000 EBH on the CWB without conducting a SLEP.

The third class of solution options involved approaches for satisfying the MCS requirement using fewer C-130s. Reducing the number of C-130s required to meet the MCS requirement offers good leverage. One option moves some C-17s from strategic missions to the intratheater role to reduce C-130 demands and adds CRAF aircraft to ensure that strategic missions are met. However, we found this option problematic for a variety of reasons. First, potential changes in the way the Army proposes to operate could result in requirements for C-130s well beyond those identified by the MCS. Further, our analysis of the number of C-130s required to meet the MCS requirement depended on several MCS assumptions that were highly favorable to a C-130/C-17/CRAF swap. As a result of the potential increased need for intratheater airlift beyond the scope of MCS and the potential fragility of the option in view of the favorable MCS assumptions, we judged this option not to be viable.

Table 9.1 summarizes results for the options that we evaluated in detail. The final line of Table 9.1 discusses SLEPs. This is a materiel solution and should be considered in future work. The SLEP option

---

[1]   For example, a 20-percent reduction in EBH accumulation on an aircraft that has four years of useful life remaining would extend the life of the aircraft by only one year.

Table 9.1
Summary of Results

FSA Option	Delay in Need for Recapitalization (years)	NPV Cost	Other Implications
Meet MCS requirement with fewer C-130s			
Shift some C-17s to theater role; backfill with CRAF in strategic airlift			Long war dominates: Not viable
Shift more AETC aircraft during peak demand	~1–2		
Reduce EBH usage rate			
Shift more training to simulators	1–2	Savings: $7 billion	
Use companion trainers	1	Cost: $6 billion	
Shift some contingency mission to other MDS	<1	Cost: $2 billion	
Increase EBH supply			
Fly aircraft beyond 45,000 EBH (fly to 56,000 EBH)	~9	Uncertain	Unacceptably high risk
SLEP	~20	TBD	

is a materiel solution that repairs, refurbishes, or replaces as necessary critical components (structural components in the CWB, for example). Our initial assessment of this option indicates that it may be a cost-effective solution and should be evaluated in a future analysis along with new aircraft options.[2]

A final observation is that a better understanding of the health of each aircraft would be highly useful and would allow better fleet management. However, these options do not delay the need for recapitalization at this point.

---

[2] The analysis in question is addressed in Kennedy et al., 2010.

In this work, we concluded that no viable nonmateriel solution or combination of nonmateriel solutions would delay the need to recapitalize the fleet by more than a few years. As a result, an AoA should be undertaken to evaluate potential materiel solutions.[3]

---

[3] In this instance, we conducted the UIAFMA (Kennedy et al., 2010). This FSA considers SLEPs and new aircraft buys to be materiel solutions and therefore defers them to the UIAFMA.

C-130 Structural Issues

This appendix first describes the evolution of the wing sections and then discusses the structural issues limiting the service life of the C-130. We also discuss sources of uncertainty, including aircraft health, aircraft tracking, and inspections.

Design Changes

Center Wing Box Design Changes

The original design of C-130E center wings experienced significant fatigue cracking after only six years of service. A redesign of that structure in 1968 reduced the maximum design stress loads on the center wing and replaced much of the corrosion- and fatigue-prone 7075-T6 aluminum alloy with a 7075-T73 alloy having better properties. The Air Force retrofitted B and E models with the redesigned structure and incorporated the design change in production aircraft.[1] These design changes enhanced service life and corrosion resistance.[2] The serial numbers of the affected aircraft (new center wing design [1968], 45,000-EBH life) are as follows:

---

[1] Later-model C-130E airplanes that received the new, more damage-tolerant CWB during production—starting with LMAC SN 4299—are sometimes referred to as the C-130E* model to distinguish them for structural integrity purposes (WR-ALC, 1995).

[2] WR-ALC, 1995; Bateman, 2005.

- C-130B/E retrofit (1968–1972) (LMAC serial numbers 3501–4298)
- C-130E*/H production (1968–) (LMAC serial numbers 4299–).

A new center wing design developed in 1991 significantly increased durability (90,000 EBH) in the severe fatigue environments AFSOC C-130s endure. USAF had this so-called special operations forces (SOF) center wing retrofitted to 79 SOF aircraft.

In the early 90s, Lockheed incorporated some of the SOF durability enhancements into a redesigned center wing for new production C-130Hs and subsequently C-130Js, beginning with serial number 5306.[3] This center wing also had a life of 45,000 EBH.[4]

Although center wings of E, H, and J air mobility aircraft differ in some details, they all essentially have similar configurations, and Lockheed assumes them to have the same basic potential service life of 45,000 EBH. Because design changes have progressively incorporated more-corrosion-resistant materials, finishes, and sealants, not all the CWBs are equally resistant to corrosion damage.[5]

Outer-Wing Design Changes

The structural configuration of the C-130's outer wings has undergone extensive design changes through the years, via modifications, retrofits, and new production. Four C-130s with the original outer-wing design have been lost, two operated by USAF, because of outer-wing structural failures.[6]

A C-130E full-scale wing fatigue test conducted from 1966 through 1970 revealed new fatigue-sensitive areas of the outer wing. At the same time, in-service airplanes, including those used in Southeast Asia, were developing fatigue and corrosion problems. This led to Engi-

---

[3]   WR-ALC, 1995; LMAC, 2005.

[4]   USAF, 2007; WR-ALC, 1995; LMAC, 2006.

[5]   LMAC, 2005; LMAC, 2006. Some caveats apply to the 45,000-EBH assessment relating to inspection requirements and the need to replace the rainbow fittings at the ends of the CWB at about 24,000 EBH. Subsequent discussion will elaborate on these points (Fraley and Christiansen, 2006).

[6]   LMAC, 2006.

neering Change Proposal (ECP) 954, Outer Wing Fatigue Preventative Modification Program, to extend the life of fatigue-sensitive areas of the outer wing. The Air Force had ECP 954 work done on the entire fleet of C-130s during depot maintenance between 1971 and 1974. A subsequent C-130B/E full-scale wing fatigue test, conducted between late 1969 and 1973, verified the satisfactory performance of the 1968 CWB design, but life enhancement of the outer wings from ECP 954 did not meet expectations. Moreover, the degree of life enhancement resulting from the ECP varied according the fatigue state of the outer wing at the time of the installation of the ECP; the timing was driven by PDM schedules and not the fatigue condition of the airplanes. Having an extended life of 60,000 EBH, this wing was incorporated on new C-130/H aircraft between 1973 and 1983, serial numbers 4542–4991.

Desiring a longer outer-wing life and building on results of fatigue testing and ECP 954 studies, the Air Force ordered a new outer-wing design that was incorporated on production aircraft starting in 1973. This design, referred to as the FY 1973 version, substituted 7075-T73 aluminum alloy for 7075-T6 in much of the outer wing, yielding more-favorable fatigue and corrosion characteristics.[7]

To provide a longer outer-wing life for B and E models, Lockheed made additional improvements to the 1973 outer-wing design in the 1980s, and the TCTO 1039 design was retrofitted to B and E models between 1984 and 1989.[8] The so-called FY 1984 outer-wing design configuration has been installed on new production Hs since 1984 and, with some minor differences, to C-130Js as well, starting with serial number 4992.[9] The incorporation of the FY 1973 and 1984 outer-wing design changes during production or by retrofit means that

---

[7] WR-ALC, 1995; Dewey Meadows, "History and Overview of Wing Improvements (Part 1)," *Lockheed Martin Service News*, Vol. 29, No. 1, 2004a.

[8] The Air Force also swapped TCTO 1039 outer wings having life remaining from retired Bs and Es and installed them on older H-model aircraft because that wing configuration had more-desirable maintenance properties (WR-ALC, 1995).

[9] Dewey Meadows, "History and Overview of Wing Improvements (Part 2)," *Lockheed Martin Service News*, Vol. 29, No. 2, 2004b.

all air mobility C-130s have a Lockheed-assessed outer-wing service life of 60,000 EBH.[10]

Fuselage Design Changes

The C-130A hydro-fatigue fuselage test conducted between 1956 and 1958 suffered three failures that would have been catastrophic had they occurred in flight. This triggered an extensive redesign of the fuselage in the late 1950s, including substitution of more-durable and corrosion-resistant aluminum alloys for skins, incorporation of fail-safe structures, use of higher-gauge skins, reductions in shell stresses, and many detailed design changes. Lockheed incorporated these changes in new production aircraft and retrofitted existing aircraft. Evolutionary changes since the original redesign have addressed service cracking and corrosion.[11] To summarize, LMAC has assessed the fuselage service life to be approximately 40,000 flight hours.[12]

Empennage Design Changes

Most changes to the empennage occurred during C-130A and B production. Service cracking has not triggered any major structural modifications or replacements. LMAC's current estimate of the empennage service life is 40,000 flight hours.[13]

All air mobility C-130s share similar service-life limits for major structural components (see Figure A.1). Of course, the fraction of service life expended for each major structural component differs from aircraft to aircraft because of differences in chronological age and usage and the fact that some aircraft have incorporated new structures through retrofits or have incorporated previously used outer-wing structures through wing swaps. The effects of the varying structural pedigrees will become apparent later as we illustrate the fleet drawdown as airplanes reach the service-life limit for each major structural component.

---

[10] USAF, 2007; WR-ALC, 1995; LMAC, 2006; Meadows, 2003, 2004a, 2004b; Ramey and Diederich, 2006.

[11] USAF, 2007; WR-ALC, 1995; LMAC, 2006.

[12] WR-ALC, 1995; LMAC, 2006.

[13] WR-ALC, 1995; LMAC, 2006.

Figure A.1
Through Redesigns and Retrofits, All U.S. Air Force C-130Es and Hs
Have the Same Structure Service-Life Potential

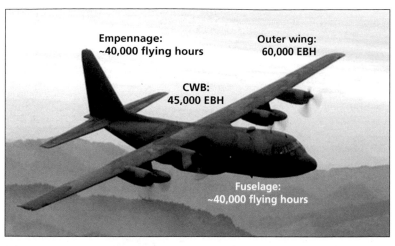

SOURCES: U.S. Air Force 2007; LMAC, 2006. Life assessments from LMAC.
RAND *MG818-A.1*

Life-Limiting Structural Issues

This section describes in more detail the nature of the fatigue damage
C-130s are experiencing, with special emphasis on the CWB, which is
the most immediate life-limiting structural component.

Center Wing

A large structural component almost 37 feet long, the CWB includes the
wing box structure, trailing edges, and internal wiring and plumbing.[14]

As illustrated in Figure A.2, fatigue cracking in the CWB typically occurs at three major load transfer points: (1) where the box
attaches to the fuselage, (2) where the inboard engine attaches to the
wing, and (3) where the outer wing attaches to the ends of the CWB.
Prior to an exhaustive TCTO 1908 inspection late in the life of the

---

[14] Dewey Meadows, "C-130 Wing Improvements Outer & Center Wings 1950s–1990s,"
presented at the 2003 Hercules Operators Conference, Atlanta, Ga., October 2003.

Figure A.2
Fatigue Cracking Is Occurring Where Fuselage, Engines, and Outer Wings Attach to Center Wing Box

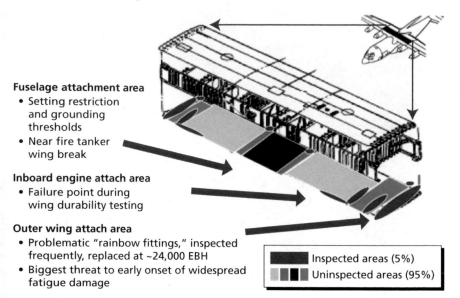

Fuselage attachment area
• Setting restriction and grounding thresholds
• Near fire tanker wing break

Inboard engine attach area
• Failure point during wing durability testing

Outer wing attach area
• Problematic "rainbow fittings," inspected frequently, replaced at ~24,000 EBH
• Biggest threat to early onset of widespread fatigue damage

Inspected areas (5%)
Uninspected areas (95%)

SOURCES: Adapted from Fraley and Christiansen, 2006; WR-ALC/LB, 2005.
RAND *MG818-A.2*

airplane, aircraft undergo inspections of small but critical areas of the lower wing surface for indications of fatigue cracking.[15]

Service events and fatigue test events underscore the criticality of the center wing structure. The fuselage attach point is the fatigue-critical area currently setting flight restriction and grounding thresholds for USAF C-130s.[16] This tracking point is about 8 inches from the point at which a center wing failed on an aging C-130A firefighting

---

[15] For example, during PDM, technicians inspect four of 140 fastener holes on each side of the CWB at wing stations 60 left and right, at the fuselage attach area. Inspections have found cracking at each of these fatigue-critical locations. About a quarter of aircraft inspected by early 2005 (44 of 184) had cracks at wing station 60. That number had grown to 71 aircraft with 102 cracks by late 2006. Inspections are continuing. (WR-ALC, 2005; Fraley and Christiansen, 2006; Christiansen, Bateman, and Navarrete, 2006.)

[16] This fatigue-critical location is usually reported as wing station 60 or 61 left and right (Fraley and Christiansen, 2006).

air tanker in 2002, resulting in the loss of the airplane and its crew.[17] While the maintenance and configuration of USAF C-130s differ in important respects from this airplane, the accident illustrates the criticality of the center wing structure and the disruptive effects of structural failures on aircraft fleets. Figure A.3 provides more details about the air tanker accident and its consequences.

The inboard engine attach point is of interest because the full-scale wing fatigue test article failed there in the last major full-scale wing fatigue test. The outer-wing attach points (the rainbow fittings) and the immediately adjacent wing panel are of interest because the

Figure A.3
Center Wing Box Failures Have Serious Consequences: Fatal C-130A Air Tanker Accident

Operator	Accident	Implications

- C-130A delivered to USAF about 1957
- 21,863 FH (estimated)
- Released to contractor from Davis-Monthan AFB in 1988
- U.S. Forest Service–contracted tanker (Hawkins & Powers Aviation Inc.)

- Occurred near Walker, California, June 17, 2002
- CWB failure (WS 53R)
- 12-in. fatigue crack hidden under doubler patch
- Multiple-site fatigue damage
- Usage and maintenance recordkeeping issues
- Maintenance issues

- Three fatalities
- Temporary and permanent aircraft groundings
- Significant loss of capability during peak fire season
- National Transportation and Safety Boardand blue-ribbon panel investigations
- 11 of 40 large contract tankers permanently grounded
- Later, in 2004, U.S. Forest Service, Department of the Interior, terminates contract for 33 large air tankers
- 16 large tankers flying in 2006 (average age 41 years)

SOURCES: Dornheim, 2002a, 2002b; Scott, 2002.
RAND *MG818-A.3*

---

[17] Fraley and Christiansen, 2006.

former does not have a fatigue life commensurate with the rest of the center wing and must be replaced at about 24,000 EBH, while the latter is a frequent unscheduled depot-level maintenance repair item.[18]

The C-130 Center Wing IRT identified the rainbow fitting as the biggest threat for early onset of widespread fatigue damage.[19] The rainbow fitting is the point at which the outer wing attaches to the CWB.[20] When cracks develop in adjacent nodes, the residual strength of the structure diminishes, and flight risks rise to unacceptable levels.[21] In-service aircraft have exhibited such cracking.[22]

The fatal in-flight breakup of a C-130A air tanker under contract to the U.S. Forest Service on June 17, 2002, and the loss of a similarly contracted PB4Y air tanker a month and a day later illustrate the stakes involved in ensuring the structural integrity of aircraft and the disruptive consequences of in-flight structural failures to aircraft fleets. A CWB failure in the C-130A resulted in the separation of both wings in flight and the consequent loss of the crew and aircraft (see Figure A.3).

Many aircraft were grounded in the aftermath of the two accidents, some permanently, others temporarily, during the peak of the fire season, and major investigations were undertaken. To this day, the Forest Service has significantly fewer large air tankers for than it did prior to the two accidents. Given the disruptive effects an accident caused by a structural failure could have on the Air Force's air mobility fleet, careful attention to C-130 structural issues seems warranted.[23]

---

[18] Fraley and Christiansen, 2006.

[19] Fraley and Christiansen, 2006.

[20] The rainbow fittings along the end of the lower wing surface refer to 13 similar structural details. In addition, there are corner fittings on both ends of the lower wing surface adjacent to the rainbow fittings. There are 11 rainbow fittings and two corner fittings along the upper wing surface, for a total of 28 attachment points between the center and outer wing (Christiansen, Bateman, and Navarrete, 2006).

[21] G. R. Bateman, "Wing Service Life Analysis Update," 2006 Hercules Operators Conference, Atlanta, Ga., October 2006.

[22] Christiansen, 2006.

[23] Michael Dornheim, "Hidden Fatigue Cracks Suspected in C-130 Crash," *Aviation Week & Space Technology*, September 2, 2002; Michael Dornheim, "Metal Fatigue Cited in Fire-fighter Crashes," *Aviation Week & Space Technology*, October 2002; William Scott, "Safety

Before showing the distribution of CWB fatigue damage across the air mobility C-130 fleet, we will illustrate and define some of the key kinds of damage experienced in CWBs and other structural components (Figure A.4).

Aircraft usually experience localized cracking earlier in their service life before exhibiting the kinds of cracking depicted in Figure A.4. Full-scale fatigue tests, when available and representative of the struc-

Figure A.4
Examples of Fatigue and Corrosion Damage in Center Wing Box

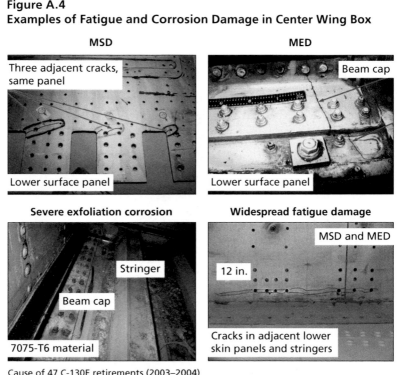

Cause of 47 C-130E retirements (2003–2004)

SOURCES: Bateman, 2005; Bateman, 2006; LMAC, 2006.

RAND MG818-A.4

Concerns Ground Aerial Firefighting Tankers," *Aviation Week & Space Technology*, December 16, 2002; Blue Ribbon Panel, *Federal Aerial Firefighting: Assessing Safety and Effectiveness*, report to the chief of the U.S. Forest Service, and to the director of the U.S. Bureau of Land Management, December 2002.

tural configurations flying, can help predict where such cracking will occur, making it easier to detect with inspections.[24]

As flight hours accumulate, cracks begin to appear in adjacent areas of a given structural element (MSD) and/or in adjacent structural elements (MED), lowering the residual strength of the structure.[25] Figure A.4 shows MSD of a single C-130 CWB wing panel as well as MED in a wing panel and adjacent beam cap. When primary fatigue cracks are present, the center wing structure has a low tolerance to MSD/MED cracks. Moreover, current inspection methods make it difficult to reliably detect small MSD/MED cracks.[26]

Ultimately, MSD and/or MED can develop to such a degree that the structure cannot meet its damage tolerance requirement. Figure A.4 illustrates an example of widespread fatigue damage where cracks have linked up in adjacent center wing lower skin panels and also exist in adjacent stringers such that the wing cannot carry design-limit loads.[27]

---

[24] LMAC uses these definitions:

> Localized cracking: Predictable locations of fatigue cracking in a principal structural element (PSE), usually determined from Full-Scale Fatigue Tests. Cracking can be reliably detected from scheduled inspections. . . . Principal Structural Element: An element of structure which contributes significantly to flight, ground, and pressurization loads, and whose integrity is essential for the maintenance of the overall structural integrity.

Definitions from G. R. Bateman, "Wing Service Life Assessment Methodology," presented at the 2004 Hercules Operators Conference, Atlanta, Ga., October 2004.

[25] LMAC uses these definitions:

> Multiple site damage (MSD): Adjacent fatigue cracks in the same principal structural element (PSE) which interact to influence residual strength. Multiple element damage (MED): Adjacent fatigue cracks in two or more principal structural elements that interact to influence residual strength. Residual strength: Capability of the structure to carry external loads in the presence of damage. (Bateman, 2004)

[26] Bateman, 2004.

[27] LMAC uses these definitions:

> Damage tolerance: Ability of the structure to retain adequate strength for a specified period in the presence of damage (fatigue cracks, corrosion, accidental damage, discrete source damage, etc.). WFD: The point in the structural life (flying hours, cycles, etc.) when MSD and/or MED cracks are of sufficient size and density such that the structure can no longer meet its Damage Tolerance Requirement (i.e., maintain required residual strength after a partial structural failure). (Bateman, 2004)

Encountering such loads in flight due to gusts or maneuvers when such damage exists can cause structural failure and loss of an aircraft. As fatigue damage accumulates, generalized cracking can occur. Under these conditions, it becomes especially difficult to detect cracking reliably during scheduled inspections.[28]

Finally, the center wing structure is also susceptible to several types of corrosion damage. Condensate can form and become trapped between fuel bladders and the structure, leaving a "waterline," such as the line on the vertical structural element on the far left of the corrosion illustration in Figure A.4.[29] Exfoliation corrosion like that shown in Figure A.4 led to the early retirement of 47 C-130Es over a recent two-year period. Repairing corrosion can involve extensive repairs and extended aircraft downtime.[30]

Outer Wing

In contrast to experience with the CWB, Lockheed has received no reports of serious cracks or corrosion on outer wings of the FY 1973–1984 design installed by retrofit or during production. These wings are, however, still relatively early in their fatigue lives.

Demonstrating a service life of 60,000 EBH, the last full-scale wing durability test article showed localized but not generalized cracking. That test used a wing test article having a configuration representative of those on operational aircraft.

Lockheed has identified several locations on the outer wings that are susceptible to MSD and MED cracking including

---

[28] LMAC uses these definitions:

Generalized cracking: Unpredictable locations of fatigue cracking in a principal structural element. [It is] difficult to detect cracking reliably from scheduled inspections. (Bateman, 2004)

[29] Gregory Shoales, Sandeep Shah, Justin Rausch, Molly Walters, Saravanan Arunachalam, and Matthew Hammond, *C-130 Center Wing Box Structural Teardown Analysis Final Report, Center for Aircraft Structural Life Extension (CAStLE)*, U.S. Air Force Academy, Colo.: Department of Engineering Mechanics, TR-2006-11, November 2006.

[30] LMAC, 2006. The C-130 ASIP Program Manager cites an example of an HC-130N (69-5824) search and rescue aircraft currently undergoing a 22-month, $1 million replacement of a corroded upper spar cap (Fraley and Christiansen, 2006).

- panels at rib attachments
- panels at engine nacelle attachment fittings
- panels at the wing joint fitting
- wing joint fittings.

Lockheed is under contract to deliver a risk assessment similar to the assessment already accomplished for the center wing by early 2008.[31] Until delivery of that analysis product, some uncertainty will exist about how risks grow as fatigue damage accumulates in the outer-wing structure.

Figure A.5 illustrates the fatigue cracking experience from the last full-scale wing fatigue test, with cumulative center wing cracks depicted by the solid line and outer-wing cracks by the dashed line. The steep slope of the CWB curve illustrates the greater prevalence of cracking in that structure compared with the outer wing.

RAND's estimate of the distribution of C-130E/H outer-wing EBH is also depicted in Figure A.5.[32] Note that even high-time outer wings on C-130E/Hs are more than 20,000 EBH from reaching the 60,000 EBH Lockheed assesses as the service life of the outer wing. It will be some time before many C-130s approach the estimated outer-wing service life demonstrated in the last test.

Fuselage and Empennage

A production redesign addressed significant problems identified in the original and only full-scale C-130A fuselage fatigue test, conducted from 1956 to 1958. Many beneficial design changes were made, although engineers retained the corrosion-prone 7075-T6 aluminum alloy for many longerons, frames, and bulkheads. Service cracking and

---

[31] LMAC, 2006.

[32] Lacking information about outer-wing EBH, the RAND team traced the history of outer-wing production, retrofits, and swaps for each outer wing and tail number using WR-ALC, 1995, and queries to the AIRCAT database. Using available data about average flying hours and severity factors, we estimated the EBH accumulated on the outer wings for each C-130. See Figure A.5 for the estimated range of outer-wing EBH for C-130Es/Hs. WR-ALC, 1995; CWB sheet, January 2007; Ramey and Diederich, 2006; and AIRCAT C-130J inventory data downloaded January 2007.

Figure A.5
Most C-130E/H Outer Wings Are Still Comparatively Early in Their Service Life

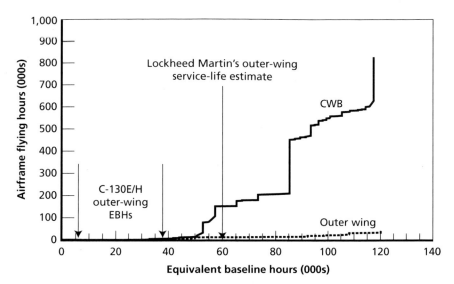

SOURCES: Adaptation from LMAC, 2006; *ASIP Master Plan*,1995; CWB sheet, January 2007; AIRCAT, 2007; Ramey and Diederich, 2006. Unofficial estimate by RAND of current outer-wing EBH uses production, replacement, or swap dates for outer wings; average SF, average flying rates.
NOTE: Thus far, there have been very few reports of significant service cracks or corrosion.
RAND *MG818-A.5*

corrosion in many locations have prompted further design refinements to C-130 fuselages, but there have been no fuselage problems of "major magnitude" since the redesign. Lockheed believes corrosion-aging processes may ultimately limit the economic service life.[33]

Canadian Forces' experience with high-time C-130E fuselages foreshadows the kinds of problems USAF may encounter as more of its aircraft pass 40,000 flight hours. Figure A.6 illustrates examples of fatigue and corrosion damage found in Canadian aircraft. The Canadi-

---

[33] LMAC, 2006; WR-ALC, 1995.

Figure A.6
Examples of Fatigue and Corrosion Damage in High-Time Canadian Forces C-130E Fuselages

Crack, aft fuselage bulkhead

Corrosion, wall of wheel well, main landing gear

Replacement of major
components accelerating

Corrosion discovery and
cost burden increasing

SOURCE: Scott, 2003.

RAND *MG818-A.6*

ans have already performed one fuselage improvement program and are planning a second program in an effort to reach 50,000 flight hours.[34]

Lockheed incorporated most design improvements to the C-130 empennage during production of A and B models. The full-scale empennage fatigue test article showed only limited cracking during testing in 1960–1961. In-service inspections have found some stress corrosion cracking of beam caps and some skin panel cracking, as well as minor surface corrosion on interior skins.[35] This has not required any major structural modifications or replacements of empennage structure.[36]

Using the service-life limit of the CWB, outer wings, and fuselage, and estimates of the flight hours or EBH accumulated by each major structural component for each individual aircraft, we can proj-

---

[34] Jason P. Scott, Spar Aerospace Ltd, "Centre Wing Damage on Canadian Forces CC130 Hercules, L-3 Communications," presented at the 2003 Hercules Operators Conference, Atlanta, Ga., October 2003.

[35] Limited access inside the empennage reportedly makes it hard to assess its true condition.

[36] LMAC, 2006; WR-ALC, 1995.

ect how inventories would decline as aircraft reach their service-life thresholds and are grounded in the absence of corrective maintenance. These projections also illustrate the order in which the Air Force would have to address successive structural issues with different parts of the airplanes to keep them flying.[37]

The 45,000-EBH CWB service life limit has the most immediate effect on C-130E/H inventories (see Figure A.7).[38] If the Air Force decides to pursue a structural SLEP, it will need to address CWB life issues first. The order in which other structural components reach their service-life limits depends on the retrofit and production history of each aircraft model. Lockheed retrofitted redesigned outer wings to C-130Es well after initial production, so C-130E outer wings are projected to reach their service-life limit after the fuselage. In contrast, many C-130Hs are flying with their original outer wings,[39] which will reach their service-life limit before the fuselage.[40]

---

[37] For purposes of illustration, the results shown in Figure A.6 assess each service-life limit independently. Obviously, the Air Force would have to address CWB issues first for aircraft to reach the outer-wing or fuselage service-life limits depicted. Aging functional systems could also require attention.

[38] Results assume all aircraft currently in the inventory undergo the maintenance needed to reach 45,000 EBH on CWBs. In fact, operating commands have elected to retire some C-130E aircraft with extensive structural damage before they reach 45,000 EBH rather than pay the high cost of repairs. So the results shown represent an optimistic assessment of aircraft inventories. We will provide more details later in this section about the population of red-X aircraft that have already been grounded before reaching 45,000 EBH. CWB sheet, July 2006; CWB sheet, January 2007; CWB sheet, March 2007.

[39] C-130H1s have retrofitted, swapped, and original production outer wings. WR-ALC, 1995; LMAC, December 2006.

[40] The study team reconstructed the history of outer-wing production, retrofits, and swaps using WR-ALC and AIRCAT data queries. The Air Force's C-130 ASIP Program Manager would not supply RAND with outer-wing EBH or severity-factor data by tail number, so the study team used average flying-hour rates and an outer-wing severity factor of 1.86 from a Lockheed Hercules Operators Conference briefing to estimate current accumulated EBH for the outer wings on each airplane. Current squadron average flying rates and the same outer-wing severity factor were used to project when outer wings would reach 60,000 EBH for each individual plane. A 2006 Mercer Engineering briefing having scatter plots of C-130 fleet outer-wing EBHs and severity factors provided a means to roughly gauge the accuracy of the RAND estimates. The average C-130 fleet outer-wing severity factor of 1.86 from the Lockheed briefing appears reasonably consistent with data in the Mercer Engineering

Figure A.7
Center Wing Box Fatigue Life Has Most Immediate Effect on C-130E/H Inventory

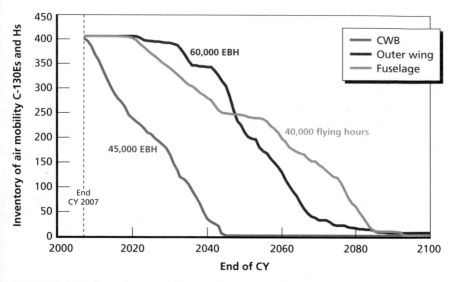

SOURCES: CWB sheet, January 2007; AIRCAT, 2007; *ASIP Master Plan*, 1995; LMAC, 2006; Ramey and Diederich, 2006.
NOTE: This figure does not include C-130J aircraft. Other figures in the document include all C-130 aircraft in the USAF inventory, including the C-130Js.
RAND *MG818-A.7*

Sources of Uncertainty

The precision implied by Figure A.7 belies considerable uncertainty about the structural health of the C-130 fleet. Many factors contribute to uncertainty about the degree of structural fatigue and corrosion damage in the C-130 fleet.

---

briefing, although the Mercer briefing shows severity factors for individual aircraft ranging from approximately 0.75 to 4.0. RAND estimates of outer-wing EBH appear just slightly greater than the C-130 fleet outer-wing EBH in the Mercer briefing. CWB sheet, January 2007; WR-ALC, 1995; LMAC, 2006; Ramey and Diederich, 2006; Brian Harper, Robert McGinty, and David Carnes, Mercer Engineering Research Center, "C-130 SOF Fleet Management Analysis," presented at the 2006 U.S. Air Force Aircraft Structural Integrity Program Conference, Memphis, Tenn., November 28–30, 2006; and AIRCAT C-130J inventory data downloaded January 2007.

Fatigue Tests and Flight Loads

Full-scale fatigue tests aid in the development of design improvements to ensure structural integrity through the life of an airplane. These tests help engineers identify fatigue-critical areas and assist in the development of inspection and maintenance procedures to deal with the fatigue cracks that appear as airplanes age. To be relevant, fatigue test articles must have a structural configuration representative of operational aircraft and must be subjected to a test load spectrum similar to loads encountered in service.

The relevance and utility of C-130 full-scale fatigue tests vary by structural component. Fuselage test results dating back to the 1950s have very little utility at this point because of all the design changes triggered by problems identified during the original fuselage test.[41] The C-130 ASIP Program Manager planned to conduct a simulated 120,000-flight-hour full-scale fatigue test of a C-130H fuselage beginning in 1997, but this test was never undertaken.[42]

Empennage and wing test results are potentially more relevant, since their structural configurations more nearly match those of current operational USAF C-130 aircraft. However, the operational loads used in the most recent wing test were measured in the mid-1980s. This adds uncertainty to projections about the service life of C-130 wing components because of changes in usage.[43]

The Royal Air Force has conducted fatigue tests more recently to assist in managing the life of its C-130Ks and C-130Js. It completed a test of its K-model wing several years ago and is about to start a new test on the C-130J wing configuration in collaboration with Australia. A full-scale fuselage fatigue test is currently under way. An ongoing C-130J operational load measurement program supports the Royal Air Force fatigue testing.[44]

---

[41] LMAC, 2006.

[42] WR-ALC, 1995.

[43] WR-ALC, 1995.

[44] Stubbs, 2005.

Operational load programs support the calculation of crack growth at critical locations by measuring aircraft usage and the severity of that usage. The last such program for the C-130 used just one instrumented aircraft in the mid-1980s, whereas ASIP standards call for instrumenting enough aircraft to achieve a 20-percent valid data capture rate for fleet usage data.[45] The lack of recent load measurements on USAF C-130Es/Hs adds to the uncertainty of EBH estimates. The Center Wing IRT recognized this and recommended implementation of a survey of loads and environment spectra to collect time histories of flight parameters to define the actual stress spectra for critical airframe areas.[46] This team recommendation had not been instituted at the time of this FSA.[47]

In contrast to the single instrumented aircraft in the mid-1980s that collected load data, ASIP standards call for instrumenting 15 percent of a fleet. The time required to establish such a program and to collect and analyze data may limit its utility for C-130Es, but comparatively younger H models could benefit.

Individual Aircraft Tracking Programs

Tracking the usage of individual aircraft and its severity is an integral part of the Air Force's current fracture-based ASIP. Given the long operational life of the C-130 fleet, there are issues associated with (1) estimates of the fatigue damage aircraft accumulated in the past, when mission records were much sparser or more aggregated than they are today; (2) gaps in reporting mission data by operational units; and (3) the accuracy of mission data submittals. The Air Force has used several different life monitoring approaches during the long operational life of Air Force C-130s (see Figure A.8). The Air Force adopted the current fracture-based methodology in the early 1980s. Over time, the

---

[45] MIL-STD-1530C, 2005.

[46] Fraley and Christiansen, 2006.

[47] Reportedly, aircraft undergoing the Avionics Modernization Program upgrade will have a wealth of data potentially available that could contribute to structural health assessments, but the Air Force will need to take affirmative actions to capture and analyze the data. Christiansen, 2006; personal communication with Hasan Ramlaoui, January 2007.

Air Force has also progressively increased the resolution with which it categorizes specific missions and the fatigue damage they cause different parts of the airplane. The "mission grid" has expanded from 9 to 1,621 missions. Fatigue damage tracking has expanded from just wing locations to other parts of the airframe.[48] The moves to a fracture-based ASIP and to progressively finer mission matrixes represent positive steps to better characterize fatigue damage but also pose a dilemma when trying to account for historical crack growth from past usage because of sparser recordkeeping in the past. To reconstruct fatigue damage incurred prior to mid-1987, the Air Force had to map quarterly flight data from nine mission categories into a larger mission matrix, developing what it called "historical makeup flights." As the mission

Figure A.8
Determining Current Equivalent Baseline Hours Required Major Reconstruction of Past Flight Data

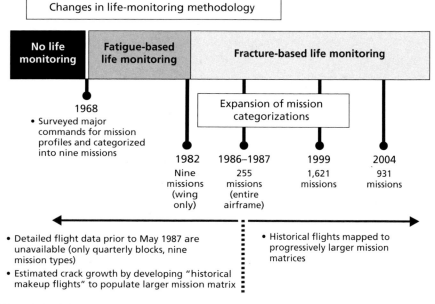

SOURCES: Christiansen, 2006; *ASIP Master Plan*, 1995; Lindenbaum, 2007.
RAND *MG818-A.8*

---

[48] WR-ALC, 1995; Lindenbaum, 2007.

matrix has progressively grown to its present size of 1,621 missions, engineers have had to map historical flights from smaller to larger mission matrices.[49]

The aforementioned process for estimating the historical accumulation of fatigue damage inevitably involves a degree of uncertainty. Moreover, even perfect reporting of mission flight parameters does not ensure accurate estimates of fatigue damage. Up-to-date data on operational loads are needed to relate flight parameters (e.g., speed at low altitudes) to loads, stresses, and crack growth.

Gaps in the reporting of flight data also introduce uncertainties in the estimation of fatigue damage. Figure A.9 illustrates how the fraction of sorties reported has varied over time. The rate of reporting has recently risen above 90 percent, but such high reporting rates have been rare in the past.[50] When flight hours are missing, the USAF's AIRCAT system uses a "window" of actual flights from the AIRCAT database to estimate the missing flight data and missing crack growth.[51]

There are also potential issues associated with the accuracy of the data that flight engineers report, although gauging this is difficult. In contrast to the automated capture of much flight data with the C-130J, C-130E/H crews manually record events during flight (e.g., time and speeds at low level, touch-and-go counts).[52] At the conclusion of flights, crews enter flight data using a web-based data-entry system, which makes rudimentary checks to catch obvious data-entry errors (e.g., out of parameter weights). The G081 Maintenance Information System also provides a means of cross checking some of the sortie data entered into AIRCAT.[53] But the data capture is anything but automatic, introducing additional uncertainty about fatigue damage estimates for individual aircraft.

---

[49] Lindenbaum, 2007.

[50] Miscellaneous queries to the AIRCAT database about outer-wing installation and removal dates and on individual aircraft reporting participation, January and March 2007.

[51] Lindenbaum, 2007.

[52] Lindenbaum, 2007.

[53] Christiansen, 2006.

Figure A.9
Degree of Reporting of Flight Data Needed for Individual Aircraft Tracking Has Varied Over Time

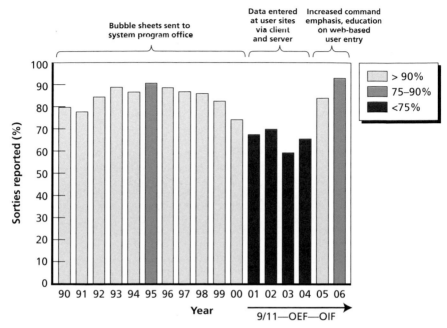

SOURCES: AIRCAT, 2007; Christiansen, 2006; Lindenbaum, 2007. C-130E-H model data from AIRCAT reports.
NOTES: Crews use paper forms to record mission data in flight. Missing flight hours filled with "normal makeup flights" to estimate missing crack growth. Flying hours cross-checked against other databases.
RAND *MG818-A.9*

Inspections

The Center Wing IRT identified deficiencies in POD and POI associated with aircraft inspections as a major Air Force–wide concern.[54] Shortcomings in POD and POI have important implications for the

---

[54] POD is defined as the "Probability that a crack will be detected using specified equipment, procedures, personnel." POI is defined as the "Probability that inspection will occur properly as specified." Fraley and Christiansen, 2006.

conduct of inspections and the viability of relying on inspections to keep high-EBH aircraft flying.[55]

Accessibility, human factors, and technology issues complicate the ability to find fatigue damage. Limited space, difficult geometries, and multilayer structures add complexity to the C-130 CWB inspection task.

Existing tools for performing inspections have serious limitations. For example, the primary TCTO 1908 inspection technique is an eddy-current surface scan that uses a probe just 1/8 inch in diameter to inspect approximately 300 ft^2 of wing surface. The Air Force additionally inspects several fatigue-critical areas of the CWB using the bolt-hole eddy-current technique. To do this inspection reliably, tapered fastener holes must be drilled out to a constant diameter. This removal of material limits the repeatability of the inspection, since oversizing the holes reduces edge clearances on some parts of the structure, increasing stress levels.[56]

Implementation of automated, large-scale techniques for C-130 CWBs has met with mixed success. An Ultra-Image ultrasonic inspec-

---

[55] An NDI tiger team concluded that the Air Force may have an institutional inspection problem. They cited principal contributors: (1) difficult locations to inspect, (2) incomplete inspection procedures, (3) inadequate oversight, (4) adequate training, (5) lack of inspector sensitivity to criticality of task, (6) human factors, and (7) newness of organizations to aircraft requirements (Michael Hatcher, WR-ALC, "Flying Through the Perfect Storm," presented at the 2005 USAF Aircraft Structural Integrity Program Conference, Memphis, Tenn., November 29–December 1, 2005). Multiple additional papers presented at the 2005 ASIP Conference provide other details about the nature of aircraft inspection problems; see CAPT John Schmidt, "Human Factors Issues in Aircraft Inspection," presented at the 2005 USAF Aircraft Structural Integrity Program Conference, Memphis, Tenn., November 29–December 1, 2005; Karl E. Kraft, Oklahoma City ALC, "NDI Coverage: Eddy Current & Ultrasound," presented at the 2005 USAF Aircraft Structural Integrity Program Conference, Memphis, Tenn., November 29–December 1, 2005; Lt Col Larry Butkus, "Undetected Cracks: Structural Significance and Root Cause Investigations," presented at the 2005 USAF Aircraft Structural Integrity Program Conference, Memphis, Tenn., November 29–December 1, 2005; John Brausch, Air Force Research Laboratory, "Addressing the NDI Crack Miss Problem in Safety of Flight Structures," presented at the 2005 ASIP Conference, Memphis, Tenn., November 29–December 1, 2005. A common theme of these papers is that inspection problems are not just a technological issue, but rather that human-factor considerations play a big role in the current limitations of inspections.

[56] Fraley and Christiansen, 2006.

tion technique was adapted to C-130 lower surface panels to inspect multiple layers and avoid the need for extensive disassembly. Although the system had previously been implemented on C-141s, it yielded inconsistent in-situ results when applied to the C-130 and was not implemented.[57]

New inspection technologies that have looked promising in laboratory environments have not performed nearly as well in real-world settings. An Air Force Academy teardown of a CWB tried an ultrasonic surface scan technique in an effort to overcome problems detecting cracks in multilayer structures, but it gave many false crack indications and exhibited only a 0.2-percent success rate in finding actual cracks.[58]

An ultraspectral ultrasonic inspection technique was implemented by the depot for forward spar caps of C-130s and exhibited results similar to the Air Force Academy experience. Automated detection software was used to interpret signals and make crack–no crack calls. Because the system flags a large number of false positives, engineers must review each positive crack indication to assess its validity.[59]

For some structural details, the critical crack length is shorter than those reliably detectable with NDI technologies. For example, the critical crack length in a rainbow fitting node at the end of the CWB is only 0.07 inches. These cracks can grow instantaneously from 0.07 to 2.5 inches. NDI can reliably detect only fractured nodes with 2.5-inch cracks.[60]

Human factors can also limit the effectiveness of inspections. Inspectors must often assume awkward positions to perform inspections. Meticulously inspecting hundreds or even thousands of fastener holes can be very tedious and requires great discipline and concentration. The Center Wing IRT found that the POI for the first ten C-130s undergoing the TCTO 1908 inspection was only about 50 percent.

---

[57] Fraley and Christiansen, 2006.

[58] Shoales et al., 2006. Other sources also cite disappointing results with new inspection technologies (Fraley and Christiansen, 2006).

[59] Fraley and Christiansen, 2006.

[60] Christiansen, Bateman, and Navarrete, 2006; Christiansen, 2006.

This led to a requirement that bolt-hole eddy-current inspections of the CWB be performed twice with independent inspectors and with engineering oversight to reduce the chances of missing cracks.[61]

Service cracking experiences and controlled tests further illustrate some of the practical limitations of inspections. As noted earlier, inspections cannot reliably detect cracks of critical length, 0.07 inches, in rainbow fitting nodes. Such cracking is reliably detectable only when cracks have grown to 2.5 inches in length. In-service aircraft have been found with several consecutive nodes cracked. This significantly reduces residual strength, resulting in aircraft flying at several orders of magnitude greater risk of structural failure than the threshold specified by MIL-STD-1530C.[62]

Test results also illustrate that inspection limitations may result in aircraft flying at higher-than-desired risks. The Air Force removed four original outer wings from C-130Es and tested them to failure. One of the three wings failed below its design load limit because of fatigue cracking. Two others failed because of fatigue cracking well below their ultimate design loads (1.5 times the design load limit). Lockheed and USAF inspections prior to the tests failed to find the fatigue cracks on two of the three wings that failed below ultimate design load.[63]

These experiences have led the C-130 ASIP Program Manager and other members of the structural community to conclude that "inspections [are] insufficient to ensure safety for aircraft with widespread fatigue damage."[64]

Even if the inspections could reliably detect the small MSD/MED cracks characteristic of high-EBH aircraft, cost and time considerations can limit the benefits of continued inspections. When generalized cracking begins to occur, inspectors cannot focus on just a few fatigue-critical locations. Larger areas must be inspected, taking aircraft out of service for extended periods for costly inspections, and

---

[61] Fraley and Christiansen, 2006.

[62] Bateman, 2006; Fraley and Christiansen, 2006.

[63] Fraley and Christiansen, 2006.

[64] Fraley and Christiansen, 2006.

inspections must be done more frequently, further increasing aircraft downtime.

Notwithstanding these limitations, the Air Force does have a process using the extensive inspection and repair protocol called out in TCTO 1908 to address fatigue risks present at 38,000 EBH and allow operation to the assessed service-life limit of the CWB of 45,000 EBH. Improvements in NDI technologies and initiatives to address human factors issues associated with inspections may reduce uncertainties about structural health and further enhance the Air Force's confidence in safely operating individual C-130s to the 45,000-EBH service-life limit.

Bibliography

Adams, T. E., and F. H. Chunn, *C-130B EAmpennage Fatigue Test Results*, Vol. I: *Simulation of 30,000 Flight Hours and 12,000 Landings*, Lockheed Aircraft Corporation, ER-5349, March 25, 1963a.

————, *C-130B Empennage Fatigue Test Results*, Vol. II: *Simulation of 60,000 Flight Hours and 24,000 Landings*, Lockheed Aircraft Corporation, ER-5349, July 22, 1963b.

AIRCAT—*See* C-130 System Program Office.

Air Force Instruction 11-2C-130, *Flying Operations*, Vol. 1: *C-130 Aircrew Training*, Washington, D.C.: Department of the Air Force, July 19, 2006.

Air Force Instruction 65-503, *U.S. Air Force Cost and Planning Factors*, Washington, D.C.: Secretary of the Air Force, February 4, 1994.

Air Force Materiel Command, *Analysis Handbook, A Guide for Performing Analysis Studies: For Analysis of Alternatives or Function Solution Analyses*, Kirtland Air Force Base, N.M.: Office of Aerospace Studies, July 2004. As of May 13, 2007: http://www.oas.kirtland.af.mil/AoAHandbook%5CAoA-Handbook.pdf

Air Mobility Command, *Airlift Mobility Planning Factors*, Scott Air Force Base, Ill.: AMC Regional Plans Branch, AFPAM 10-1403, December 2003.

————, *Aircrew/Aircraft Tasking System (AATS)*, Scott Air Force Base, Ill., November 16, 2005.

————, *Migrating Mobility Forces into AEF*, Scott Air Force Base, Ill., December 2006.

Air Mobility Support Engineering, "C-130 Center Wing: Service Life Issues," *Lockheed Martin Service News*, Vol. 30, No. 2, 2005a, pp. 3–4.

————, "C-130 Center Wing: History & Overview," *Lockheed Martin Service News*, Vol. 30, No. 2, 2005b, pp. 5–6.

ASIP Master Plan—*See* WR-ALC, 1995

————, "C-130 Center Wing: Replacement Program," *Lockheed Martin Service News*, Vol. 30, No. 2, 2005c, pp. 7–8.

Baig, Asad, SPAR Aerospace Limited, "Royal New Zealand Air Force C-130 Life Extension Program," presented at the 2005 Hercules Operators Conference, Atlanta, Ga., October 2005.

Bateman, G. R., "Wing Service Life Assessment Methodology," presented at the 2004 Hercules Operators Conference, Atlanta, Ga., October 2004.

————, "Wing Service Life Assessment Methodology & Results," presented at the 2005 Hercules Operators Conference, Atlanta, Ga., October 2005.

————, "Wing Service Life Analysis Update," presented at the 2006 Hercules Operators Conference, Atlanta, Ga., October 2006.

Bateman, G. R., and P. Christiansen, "C-130 Center Wing Fatigue Cracking, A Risk Management Approach," presented at the 2005 U.S. Air Force Aircraft Structural Integrity Program Conference, Memphis, Tenn., November 29–December 1, 2005.

Blue Ribbon Panel, *Federal Aerial Firefighting: Assessing Safety and Effectiveness*, report to the chief of the U.S. Forest Service and the director of the U.S. Bureau of Land Management, December 2002.

Boito, Michael, Gregory G. Hildebrant, Fred Timson, *Analysis of Alternatives (AOS) for KC-135 Recapitalization: Appendix F—Operating and Support Costs*, Santa Monica, Calif.: RAND Corporation 2005, Not Available to the General Public.

Brausch, John, Air Force Research Laboratory, "Addressing the NDI Crack Miss Problem in Safety of Flight Structures," presented at the 2005 USAF Aircraft Structural Integrity Program Conference, Memphis, Tenn., November 29–December 1, 2005.

Bush, George W., *The National Security Strategy of the United States of America*, Washington, D.C.: The White House, September 2002.

Butkus, Larry, Lt Col, U.S. Air Force, "Undetected Cracks: Structural Significance and Root Cause Investigations," presented at the 2005 USAF Aircraft Structural Integrity Program Conference, Memphis, Tenn., November 29–December 1, 2005.

C-130 System Program Office, Automated Inspection, Repair, Corrosion and Aircraft Tracking (AIRCAT) database queries, various dates.

————, C-130 center wing box data spreadsheets from Automated Inspection, Repair, Corrosion and Aircraft Tracking database, July 12, 2006.

————, C-130 AIRCAT Center Wing Equivalent Baseline Hours (EBH) Report, spreadsheet, Robins Air Force Base, Ga.: U.S. Air Force Materiel Command, January 3, 2007.

Cardinal, Joseph W., and Hal Burnside, Southwest Research Institute, "Damage Tolerance Risk Assessment of T-38 Wing Skin Cracks," presented at the 2005 U.S. Air Force Aircraft Structural Integrity Program Conference, Memphis, Tenn., November 29–December 1, 2005.

Chairman of the Joint Chiefs of Staff Instruction (CJCSI) 3170.01E, *Joint Capabilities Integration and Development System (JCIDS)*, Washington, D.C., May 11, 2005.

Chilcott, Andrew, Structural Monitoring Systems, Ltd, "Comparative Vacuum Monitoring CVM™," presented at the 2006 Hercules Operators Conference, Atlanta, Ga., October 2006.

Christiansen, Peter, Warner Robins Air Logistics Center, "Assessment of U.S. Air Force Center Wing Cracking," presented at the 2006 Hercules Operators Conference, Atlanta, Ga., October 2006.

Christiansen, P., G. R. Bateman, and A. Navarrete, "C-130 Center Wing MSD/MED Risk Analysis," presented at the 2006 U.S. Air Force Aircraft Structural Integrity Program Conference, Memphis, Tenn., November 28–30, 2006.

Chunn, F. H., and T. E. Adams, *C-130B Empennage Fatigue Test Results*, Vol. III: *Simulation of 84,000 Flight Hours and 33,600 Landings*, Lockheed Aircraft Corporation, ER-5349, August 9, 1963.

Davis, H. Don, and Dave Schmidt, Av-Dec, "Corrosion Prevention Using Polyurethane Sealants," presented at the 2006 Hercules Operators Conference, Atlanta, Ga., October 2006.

Department of the Air Force, Contractor Logistics Support Brochures, various programs, prepared for FYs 2006, 2007, and 2008.

DoD—*See* U.S. Department of Defense.

Dornheim, Michael, "Hidden Fatigue Cracks Suspected in C-130 Crash," *Aviation Week & Space Technology*, September 2, 2002.

————, "Metal Fatigue Cited in Firefighter Crashes," *Aviation Week & Space Technology*, October 2002.

Dragan, Krzysztof, Capt, Polish Air Force, Slawomir Klimaszewski, Lt Col, Polish Air Force, Andrzej Leski, Maj, Polish Air Force, Air Force Institute of Technology, Warsaw, Poland, "NDE Approach for Structural Integrity Evaluation of MIG-29 Aircraft," presented at the 2005 U.S. Air Force Aircraft Structural Integrity Program Conference, Memphis, Tenn., November 29–December 1, 2005.

Farrar, Charles R., and Nick A. J. Lieven, "Damage Prognosis: The Future of Structural Health Monitoring," *Philosophical Transactions of the Royal Society A*, December 12, 2006.

Fontoura, Sid, Spar Aerospace Ltd, "U.S. Air Force C-130 CW Teardown Results, L-3 Communications," presented at the 2006 Hercules Operators Conference, Atlanta, Ga., October 2006.

Fraley, Marian, "C-130 Center Wing Status," Robins Air Force Base, Ga.: WR-ALC/LB, February 16, 2005.

Fraley, Marian, and Peter Christiansen, "C-130 Groundings and Restrictions, Warner Robins Air Logistics Center, 330 ACSG," briefing to RAND, U.S. Air Force, Program Analysis and Evaluation, Rosslyn, Va., October 20, 2006.

General Accounting Office [now the Government Accountability Office], *Financial Management: DoD's Liability for Aircraft Disposal Can Be Estimated*, Washington, D.C., GAO/AIMD-98-9, November 1997.

Harper, Brian, Robert McGinty, and David Carnes, Mercer Engineering Research Center, "C-130 SOF Fleet Management Analysis," presented at the 2006 U.S. Air Force Aircraft Structural Integrity Program Conference, Memphis, Tenn., November 28–30, 2006.

Hatcher, Michael, Warner Robins Air Logistics Center, "Flying Through the Perfect Storm," presented at the 2005 USAF Aircraft Structural Integrity Program Conference, Memphis, Tenn., November 29–December 1, 2005.

Hawker Beechcraft Corporation, Beech King Air 350 photo, undated.

Hill, Lewis D., Doris Cook, and Aron Pinker, *Gulf War Air Power Survey*, Vol. 5, Pt. I: *A Statistical Compendium*, Washington, D.C.: U.S. Government Printing Office, 1993. As of July 13, 2009: http://www.airforcehistory.hq.af.mil/Publications/Annotations/gwaps.htm

Joint Publication 4-01.6, *Joint Logistics Over the Shore (JLOTS)*, Appendix B, Washington, D.C.: Joint Chiefs of Staff, August 2005.

Kennedy, Michael, David T. Orletsky, Anthony D. Rosello, Sean Bednarz, Katherine Comanor, Paul Dreyer, Chris Fitzmartin, Ken Munson, William Stanley, and Fred Timson, *USAF Intratheater Airlift Fleet Mix Analysis*, Santa Monica, Calif.: RAND Corporation, 2010, Not Available to the General Public.

Kraft, Karl E., Oklahoma City Air Logistics Center, "NDI Coverage: Eddy Current & Ultrasound," presented at the 2005 USAF Aircraft Structural Integrity Program Conference, Memphis, Tenn., November 29–December 1, 2005.

Leeuw, Ruud, "Tanker 130 Down!!!" web page, July 11, 2003. As of April 2, 2007: http://www.ruudleeuw.com/tanker130.htm

Lindenbaum, J. A., *Equivalent Baseline Hours Methodology of U.S. Air Force C-130E-H Individual Aircraft Tracking Program (IATP), Automated Inspection, Repair, Corrosion and Aircraft Tracking (AIRCAT)*, Atlanta, Ga.: Lockheed Martin Aeronautics Company, LG07ER0221, February 12, 2007.

Liu, Ko-Wei, *C-130 Center Wing Box Wing Life Enhancement*, Long Beach, Calif.: The Boeing Company, January 2007.

LMAC—*See* Lockheed Martin Aeronautics Company.

Lockheed Martin Aeronautics Company, Fleet Viability Board briefings, December 6, 2006.

MCS—*See* U.S. Department of Defense and Joint Chiefs of Staff.

Meadows, Dewey, "C-130 Wing Improvements Outer & Center Wings 1950s–1990s," presented at the 2003 Hercules Operators Conference, Atlanta, Ga., October 2003.

———, "History and Overview of Wing Improvements (Part 1)," *Lockheed Martin Service News*, Vol. 29, No. 1, 2004a, pp. 4–5.

———, "History and Overview of Wing Improvements (Part 2)," *Lockheed Martin Service News*, Vol. 29, No. 2, 2004b, pp. 11–12.

Military Standard 1530C (MIL-STD-1530C), *Aircraft Structural Integrity Program (ASIP)*, Wright-Patterson Air Force Base, Ohio: Aeronautical Systems Center, November 1, 2005.

O'Connor, Seamus, "'7-Day Option' Gets Second Look," *Air Force Times*, June 6, 2008. As of November 23, 2009: http://www.airforcetimes.com/news/2008/06/airforce_7dayoption_060408/

Office of Management and Budget, "Discount Rates for Cost-Effectiveness, Lease Purchase, and Related Analyses," rev., Washington, D.C.: Executive Office of the President, OMB Circular No. A-94, App. C, January 2007. As of May 13, 2007: http://www.whitehouse.gov/omb/circulars/a094/a94_appx-c.html

Orletsky, David T., Anthony D. Rosello, and John Stillion, *Intratheater Airlift Functional Area Analysis (FAA), Santa Monica*, Calif.: RAND Corporation, MG-685-AF, 2011. As of February 3, 2011: http://www.rand.org/pubs/monographs/MG685.html

Ramey, S. F., and J. C. Diederich, "Operational Usage Evaluation and Service Life Assessment," presented at the 2006 Hercules Operators Conference, Atlanta, Ga., October 2006.

Reid, Len, Fatigue Technology Inc., "Aging Aircraft Repair Strategies Utilizing Cold Expansion Technology," presented at the 2005 USAF Aircraft Structural Integrity Program Conference, Memphis, Tenn., November 29–December 1, 2005.

Sandia National Laboratories, "Sensors May Monitor Aircraft for Defects Continuously, Structural Health Monitoring Systems Accepted by Boeing, Validated by Airlines," news release, Albuquerque, N.M., July 18, 2007.

Schmidt, John, CAPT, "Human Factors Issues in Aircraft Inspection," presented at the 2005 USAF Aircraft Structural Integrity Program Conference, Memphis, Tenn., November 29–December 1, 2005.

Scott, Jason P., Spar Aerospace Ltd, "Centre Wing Damage on Canadian Forces CC130 Hercules, L-3 Communications," presented at the 2003 Hercules Operators Conference, Atlanta, Ga., October 2003.

Scott, William, "Safety Concerns Ground Aerial Firefighting Tankers," *Aviation Week & Space Technology*, December 16, 2002.

Shoales, Gregory, Sandeep Shah, Justin Rausch, Molly Walters, Saravanan Arunachalam, and Matthew Hammond, *C-130 Center Wing Box Structural Teardown Analysis Final Report*, U.S. Air Force Academy, Colo.: Department of Engineering Mechanics, Center for Aircraft Structural Life Extension (CAStLE), TR-2006-11, November 2006.

Speckman, Holger, and Rudolf Henrich, *Structural Health Monitoring (SHM): Overview of Airbus Activities,* presented at the 16th World Conference on NDT, Montreal, Canada, August 30–September 3, 2004.

Stillion, John, David T. Orletsky, and Anthony D. Rosello, *Intratheater Airlift Functional Needs Analysis (FNA)*, Santa Monica, Calif.: RAND Corporation, MG-822-AF, 2011. As of February 3, 2011:
http://www.rand.org/pubs/monographs/MG822.html

Stubbs, Mark, Squadron Leader, Royal Air Force, "RAF Structures Programmes and Fatigue Management," presented at the 2006 Hercules Operators Conference, Atlanta, Ga., October 2005.

Time Compliance Technical Order 1908, "Inspection of C-130 Center Wing for Generalized Cracking," June 1, 2006.

Time Compliance Technical Order 1039, Retrofit Installation of Redesigned Outer Wing, C-130 Aircraft, October 1, 1983.

TRADOC Analysis Center—*See* U.S. Army Training and Doctrine Command Analysis Center.

U.S. Air Force, official photos, various sources, 2007.

Air Force Financial Management and Comptroller, *United States Air Force, Committee Staff Procurement Backup Book: FY 2007 Budget Estimates, Aircraft Procurement, Air Force*, Vol. I, Washington, D.C.: U.S. Air Force, February 2006. As of November 23, 2009:
http://www.saffm.hq.af.mil/shared/media/document/AFD-070214-050.pdf

————, *Committee Staff Procurement Backup Book: FY 2008/2009 Budget Estimates, Aircraft Procurement, Air Force*, Vol. I, Washington, D.C., U.S. Air Force, February 2007. As of November 23, 2009:
http://www.saffm.hq.af.mil/shared/media/document/AFD-070212-004.pdf

U.S. Army Aviation Center, Futures Development Division, Directorate of Combat Developments, *Army Fixed Wing Aviation Functional Area Analysis Report*, Fort Rucker, Ala., June 3, 2003a.

————, *Army Fixed Wing Aviation Functional Needs Analysis Report*, Fort Rucker, Ala., June 23, 2003b.

————, *Army Fixed Wing Aviation Functional Solution Analysis Report*, Fort Rucker, Ala., June 8, 2004.

U.S. Army Training and Doctrine Command 3170 Analysis Center, *Joint Cargo Aircraft/Future Cargo Aircraft Analysis of Alternatives*, final results scripted brief, Fort Leavenworth, Kan., TRAC-F-TR-07-027, March 2005a.

————, *Future Cargo Aircraft (FCA) Analysis of Alternatives (AoA)*, Fort Leavenworth, Kan., TRAC-TR0-5-18, July 18, 2005b, Not Available to the General Public.

U.S. Army and U.S. Air Force, "Way Ahead for Convergence of Complementary Capabilities," memorandum of understanding, February 2006.

U.S. Central Air Force, Assessment and Analysis Division, *Operation IRAQI FREEDOM—By the Numbers*, Shaw AFB, S.C., April 2003.

U.S. Code, Title 10, Subtitle E, Pt. II, Ch. 1209, Sec. 12302.

U.S. Department of Defense and the Joint Chiefs of Staff, *Mobility Capabilities Study*, Washington, D.C., December 2005, Not Available to the General Public.

————, *Quadrennial Defense Review Report*, Washington, D.C., February 6, 2006.

————, "Unified Facilities Criteria (UFC): DoD Facilities Pricing Guide," Vers. 7, March 7, 2007.

————, *The National Defense Strategy of the United States of America*, Washington, D.C., June 2008.

————, *Mobility Capabilities & Requirements Study 2016*, Washington, D.C., February 26, 2010, Not Available to the General Public.

Vaughan, Kevin, "Colorado Lawmakers Worry About Wildfire Readiness," *Rocky Mountain News*, 2006. As of April 2, 2007:
http://www.firerescue1.com/print.asp?act=print&vid=103980

Warner Robins Air Logistics Center, *Aircraft Structural Integrity Program Master Plan for C-130 Aircraft*, Robins Air Force Base, Ga.: U.S. Air Force Materiel Command, December 1995.

———, "C-130 Center Wing Status," briefing, Robins Air Force Base, Ga.: U.S. Air Force Materiel Command, February 9, 2005.

WR-ALC—*See* Warner Robins Air Logistics Center.